Supercapacitors and Their Applications

Owing to their high power density, long life, and environmental compatibility, supercapacitors are emerging as one of the promising storage technologies, but with challenges around energy and power requirements for specific applications. This book focuses on supercapacitors including details on classification, charge storage mechanisms, related kinetics, and thermodynamics. Materials used as electrodes, electrolytes, and separators, procedures followed, characterization methods, and modeling are covered, along with emphasis on related applications.

Features:

- Provides an in-depth look at supercapacitors, including their working concepts and design
- Reviews detailed explanation of various characterization and modeling techniques
- Gives special focus to the application of supercapacitors in major areas of environmental as well as social importance
- Covers cyclic voltammetry, charging–discharging curves, and electrochemical impedance spectroscopy as characterization techniques
- Includes a detailed chapter on historical perspectives on the evolution of supercapacitors

This book is aimed at researchers and graduate students in materials science and engineering, nanotechnology, chemistry in batteries, and physics.

Supercapacitors and Their Applications

Fundamentals, Current Trends, and Future Perspectives

Edited by
Anjali Paravannoor and
Baiju K. V.

CRC Press
Taylor & Francis Group
Boca Raton London New York

CRC Press is an imprint of the
Taylor & Francis Group, an **informa** business

First edition published 2023
by CRC Press
6000 Broken Sound Parkway NW, Suite 300, Boca Raton, FL 33487-2742

and by CRC Press
4 Park Square, Milton Park, Abingdon, Oxon, OX14 4RN

CRC Press is an imprint of Taylor & Francis Group, LLC

ISBN: 9781032192604 (hbk)
ISBN: 9781032192628 (pbk)
ISBN: 9781003258384 (ebk)

DOI: 10.1201/9781003258384

Typeset in Times
by Deanta Global Publishing Services, Chennai, India

Contents

Editor Biographies

Dr Anjali Paravannoor is currently working as a DST-INSPIRE faculty at Kannur University, Kerala, India. She earned her PhD degree (2015) in Nanomaterials from Amrita Viswavidyapeetham University, Ettimadai, India. She received her master's degree in Nanoscience and Technology from Bharathiar University, Coimbatore, India. Over the last ten years, she has been actively involved in research in the field of supercapacitor systems. She has more than 30 publications in peer-reviewed international journals and several book chapters in the area of energy generation and storage. Currently, she is working on 2D nanostructures for pseudocapacitor applications. Her research interests also include the development of novel anode materials for lithium-ion and sodium-ion battery applications.

Dr Baiju Kizhakkekilikoodayil Vijayan is currently working as an Assistant Professor at Kannur University, Kerala, India. He has worked as a Scientist at the Centre for Materials for Electronics Technology, Thrissur, India, which is a research and development laboratory under the Ministry of Electronics and Information Technology, Government of India, from 2011 to 2016. He also worked as a Post-Doctoral Associate in North-Western University, Evanston, Illinois, USA, from October 2008 to April 2011, in the area of photocatalytic reduction of carbon dioxide to fuel. He earned his PhD from the National Institute of Interdisciplinary Science and Technology, formerly known as RRL Trivandrum, India, on the topic of sol-gel synthesis of nanomaterials in 2008. To his credit he has more than 60 published papers (>3500 citations) in peer-reviewed international journals and 7 patents. He is currently working in the area of nanostructured materials for catalyst, photocatalyst, supercapacitor, and solar cell applications.

Contributors

Shyamili Ashok
Department of Chemistry
Nirmalagiri College
Kannur, India

Margandan Bhagiyalakshmi
Department of Chemistry
Central University of Kerala
Kerala, India

Mijun Chandran
Department of Chemistry
Central University of Tamil Nadu
Thiruvarur, India

Gopakumar G.
Department of Chemistry
University of Kerala
Kollam, India

Fabeena Jahan
School of Chemical Sciences
Kannur University
Payyanur, India

Sarayu Jayadevan
Department of Chemistry
Sir Syed College
Taliparamba, India

Deepu Joseph
Department of Physics
Nirmalagiri College
Kannur, India

Vijila Kalarivalappil
Basic Science and Humanities
    Division
Indian Naval Academy
Kannur, India.

Kunnambeth M. Thulasi
School of Chemical Sciences
Kannur University
Payyanur, India

Shidhin M.
Central European Institute of
    Technology
Brno University of Technology
Brno, Czech Republic

Sreejesh M.
Centre for Nano and Soft Matter
    Sciences (CeNS)
Bangalore, India

Sindhu Thalappan Manikkoth
School of Chemical Sciences
Kannur University
Payyanur, India

Sruthi Maruthiyottu Veettil
Central NMR Facility
CSIR – National Chemical Laboratory
Pune, India

Shridhar Mundinamani
Department of Physics
Siddaganga Institute of Technology
Tumakuru, India

Shilpa M. P.
Department of Physics,
Manipal Institute of Technology
Manipal, India

Nijisha P.
School of Chemical Sciences
Kannur University
Payyanur, India

Deepthi Panoth
School of Chemical Sciences
Kannur University
Payyanur, India

Anjali Paravannoor
School of Chemical Sciences
Kannur University, India

Divya Puthussery
Department of Nanoscience and
    Technology
Calicut University
Kerala, India

Asha Raveendran
Department of Chemistry
National Institute of
    Technology
Puducherry, India

Gurumurthy S. C.
Department of Chemistry
Manipal Institute of
    Technology
Manipal, India

Sudhakar Y. N.
Department of Chemistry
Manipal Institute of
    Technology
Manipal, India

S. Anas
Department of Chemistry
University of Kerala
Kollam, India

Shivakumar Shetty
Department of Physics
Manipal Institute of Technology
Manipal, India

Jasmine Thomas
Department of Chemistry
Nirmalagiri College
Kannur, India

Nygil Thomas
Department of Chemistry
Nirmalagiri College
Kannur, India

Sujith K. V.
Department of Chemistry
Payyanur College
Payyanur, India

Baiju Kizhakkekilikoodayil Vijayan
School of Chemical Sciences
Kannur University
Payyanur, India

Mari Vinoba
Kuwait Institute for Scientific Research
Kuwait

Preface

The availability of energy sources remains one of the major challenges across the globe, particularly in the perspective of sustainable availability, toxicity, low cost, and environmental impact. Research is being focused more and more towards the development of new and renewable energy sources like the sun, wind, and tide, again leaving an ever-increasing requirement for efficient energy storage devices. The most prominent technologies of choice are batteries; however, during the past few decades, supercapacitors have been emerged as a promising alternative, owing to their high power density, cyclic stability, and environmental friendliness. The compilation presented in this book attempts to focus on the various aspects of supercapacitor technology, ranging from their principle and design to applications. The book focuses mainly on two major aspects of supercapacitors: fundamentals and their specific applications. In spite of the various advanced scientific concepts developed for the design of highly efficient supercapacitor systems, there are still lacunae to filled in the comprehensive literature on the topic, mainly in the perspective of broad application areas. Hence the book tries to capture all the available knowledge starting from the historical background to latest applications, taking into account the commercial as well as academic interests of the readers. Significant basic concepts of supercapacitors such as their classification; charge storage mechanisms; kinetics and thermodynamics of the processes; materials used as electrode, electrolyte, and separators; and procedures followed are covered in the initial part. This will especially help the academicians and researchers in rationalizing the materials and designs that can be taken up further. Potential material characterization methods and modeling are also covered in the book in a simple way which makes reading interesting for beginners. The latter part of the book discusses the future directions and prospects of supercapacitor technology, starting from the manufacture of industrial supercapacitors extending to specific applications like electric vehicles, solar cells, wind turbines, AC line filtering, and load leveling. Advanced supercapacitor applications in the biomedical, military, and aerospace sectors are also provided in this section. This application-bound perspective of the book makes it suitable for scientists, industrialists, and students. Moreover, the conceptual framing of this book makes it appropriate for inclusion in the curriculum of various graduate and postgraduate courses in basic science as well as engineering.

1 Historical Perspectives

Vijila Kalarivalappil and
Baiju Kizhakkekilikoodayil Vijayan

CONTENTS

1.1 INTRODUCTION

Energy consumption has become an inevitable factor to achieve economic growth and hence human development. Nowadays, the depletion of conventional fuels has triggered the search for alternative energy resources competing with traditional sources like coal, oil, and gas. Environmental issues including global warming have also made current innovative technologies focus on renewable energy resources [1]. The different renewable energy sources that have fewer adverse effects on our environment include solar energy, wind power, tidal power, geothermal energy, etc. As a part of this clean energy portfolio, the demand for eco-friendly, high-performance energy storage devices also increases dramatically [2–6]. A capacitor is one such electrical component in a circuit, which is used to store charge. It works on the principle of an increase in the capacitance of a conductor when another earthed conductor is brought near it. Hence, a capacitor consists of two equal and oppositely

DOI: 10.1201/9781003258384-1

1

charged plates separated by a distance and its ability to store charge is known as capacitance. Capacitors are mainly used in electronic circuits to charge or discharge electricity, which is applied to back up the circuits of microcomputers, timer circuits, and smoothing circuits of power supplies, and is also used to block the flow of DC which is applied to filter applications for the elimination of particular frequencies. Thus, the applications of capacitors range from tuning or filter circuits to hybrid electric vehicles and satellites. The advancement in nanoscience and technology have influenced electrical energy storage technologies tremendously, including batteries and capacitors [7–14]. The history of capacitance technology started in 1745, when an early version of a capacitor was invented by Ewald G. von Kleist while accidentally giving an electric charge to a medicine bottle. Later Pieter van Musschenbroek also came up with the same invention while living the city of Leiden. The Leyden jar, a primitive model of a capacitor, consists of a glass vessel with metal foils with the jar forming the dielectric of the capacitor and the metal foils forming the electrodes. During the charging process, opposite charges are accumulated on the metal electrode surface and the discharging of these charges occurs when a metal wire is connected across the metal electrodes. The capacitor technology has undergone many modifications ever since and this chapter describes the mechanisms involved in the electric capacitor, different types of capacitors, and the recent advancements in capacitor technology.

1.2   CAPACITANCE OF A CAPACITOR

The capacitance (C) of a capacitor, by definition, is the amount of charge stored per unit of voltage across its plates and it is a measure of the capacitor's ability to store charge. That is, capacitors with higher charge per unit of voltage have greater capacitance, as given by the following formula:

$$C = \frac{q}{V}$$

where q is the amount of charge on either conductor and V is the potential difference between the conductors, or in other words, it is the ratio of the amount of charge (q) to the potential difference (V).

The unit of capacity depends on the units of charge and potential used. The SI unit of capacity is the farad. One farad is the capacity of a conductor in which the potential of the conductor rises by one volt if one coulomb of charge is given. Since the practical unit of capacity is too large, much smaller units like millifarad (10^{-3} F), microfarad (10^{-6} F), nanofarad (10^{-9} F), and picofarad (10^{-12} F) are used in normal practice.

1.3   WORKING PRINCIPLE OF A CAPACITOR

To understand the charging and discharging mechanism of capacitors we will consider the most common parallel plate capacitor in which the space between the conducting plates is filled with a dielectric material. The dielectric material can be a vacuum or an insulating material with a polarizable property. For a given voltage, the dielectric constant of material gives the amount of electrical energy stored in it. The capacitors, having materials with higher dielectric constants, store higher electrical energy. The dielectric strength of a material can determine the maximum voltage that can be applied to the capacitor before its breakdown. When the battery is connected, current flows and the potential difference across the capacitor plates begins to rise, as the free electrons from the neutral atoms of the plate are attracted towards the positive potential applied to plate I which creates an excess of positive ions on plate I. Then these free electrons are pushed towards the plate II by the chemical action occurring in the battery and also with the repulsion by the negative battery potential creating a negative charge on plate II of the capacitor. A potential difference between two plates arises due to the charges developed on them and it is defined as voltage. There will not be any current flow when the potential difference across the capacitor becomes equal to that of the applied battery voltage. At this moment the capacitor is said to be fully charged and the time span taken for this is known as the charging time of the capacitor. The charging time depends upon factors such as circuit resistance and the capacitance of the capacitor.

The two plates of the capacitor hold the opposite charge for a certain time when the battery is removed. From this moment onwards it can act as a source of electrical energy. If these plates are connected to a load, the current flows through the load for a short period from the plate which is negatively charged to the plate which is positively charged until all the charge is dissipated from both plates. The time taken for this process is known as the discharging time.

C is given by the formula $C = K \dfrac{\varepsilon_0 A}{d}$ which is the capacitance of a parallel plate capacitor

Where:

- The permittivity of space ($8.85 * 10^{-12}$ F/m) is denoted by ε_0
- The relative permittivity of dielectric material denoted by K
- The separation between the plates is denoted by d
- The area of plates is denoted by A

From the above equation, it is observed that the capacitance (C) increases with the permittivity of the dielectric material, the area (A) of the plates, and decreases with the plate separation (d) [15].

1.4 ENERGY STORED IN CAPACITOR

When the battery is removed the two plates of the capacitor hold the opposite charge for a certain time, maintaining the potential difference between them. The charge is stored due to the presence of dielectric or insulator material which prevents the movement of electrons from plate II to plate I. If we assume that a charge (q) is being transferred from one plate of a capacitor to the other in an instant (V), the potential difference between the plates at that instant can be calculated as $\frac{q}{C}$. The work required to bring the total capacitor charge up to a final value q is:

$$W = \int dw = \frac{1}{c} \int_0^q q \, dq = \frac{q^2}{2C}$$

This work is stored as potential energy U in the capacitor, so that $U = \frac{q^2}{2C}$. We can also write this as $U = \frac{1}{2}CV^2$ [16].

1.5   TYPES OF CAPACITORS

There are a wide variety of capacitors depending on the practical application. The main categories are variable capacitors and non-variable capacitors. The capacitance value of variable capacitors can be changed mechanically or electronically, as the name suggests, whereas a non-variable capacitor gives fixed capacitance. Capacitors are mainly classified into three categories, namely electrostatic capacitors, electrolytic capacitors, and electrochemical capacitors

1.5.1   ELECTROSTATIC CAPACITOR

Typical electrostatic capacitors consist of two thin strips of metal foil as electrodes separated by an insulator material act as a dielectric (Figure 1.1(a)). Various types of electrostatic capacitors are available in the market commercially. The names of those capacitors originate from the type of dielectric material used for their construction. Examples of various electrostatic capacitors include air, ceramic, mica, and plastic/paper film capacitors.

1.5.1.1   Air Capacitor

In an air capacitor, the air between the plates acts as the dielectric material. The capacitance of the air capacitor can be tuned by meshing and unmeshing the capacitor

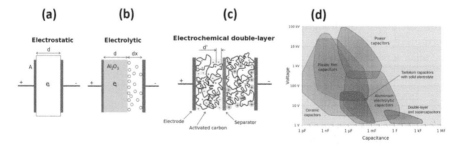

FIGURE 1.1   (a), (b), and (c) Schematic representation of the different classification of capacitors. (d) Capacitance versus voltage graph of the different class of capacitors [17, 18].

plates. So variable and fixed types of air capacitors can be made for different applications [19]. The construction of these capacitors is simple and usually consists of two sets of plates, one fixed and the other movable. The movable set of plates move in between the fixed plates so that different amounts of air can be included within the gap. As the overlap between the two sets of plates becomes larger, the capacitance becomes higher. Variable air capacitors find wide application in places where variable capacitance is required; for example, in resonant circuits like antenna impedance-matching applications or frequency mixers, radio tuners, and also for modeling electronic circuits. Most of these applications demand high power, high frequency, and low loss properties. The other applications of air capacitors include military applications which utilize their intrinsic resistance to electromagnetic pulses that could be created intentionally to destruct electronic equipment or scanners. The inferior dielectric strength of air makes air capacitors unsuitable for high voltages. The capacitance range of air capacitors lies between 100 pF and 1 nF.

1.5.1.2 Ceramic Capacitor

A ceramic capacitor is a capacitor that uses a ceramic material as the dielectric. The two most common types of ceramic capacitors widely used in electronics include multi-layer ceramic capacitors (MLCC) and ceramic disc capacitors. The capacitance value of these capacitors normally lies between 1 nF and 1 μF with low maximum rated voltage compared with electrolytic capacitors. The ceramic capacitors available today are divided into two groups: class 1 and class 2. Paraelectric dielectric materials such as rutile phase TiO_2 and perovskite titanates, along with additives of Zn, Mg, or Ta forms class 1 ceramic capacitors [20, 21]. The capacitance of these types of capacitors shows minimum change or drift with respect to temperature, frequency, and applied voltage, and they are highly preferred for applications that demand high stability and low loss. Whereas ferroelectric materials such as barium titanate along with additives like silicates of aluminum, magnesium, and aluminum oxide form class 2 capacitors [22–25]. Since the dielectric materials of class 2 ceramic capacitors are associated with high permittivity, they are suitable for less sensitive applications like decoupling and coupling applications. MLCC devices offer a significant advancement in most modern printed circuit boards (PCBs), where high component packing densities are required. A voltage range of 2–100 kV can be made to withstand power by ceramic capacitors. But smaller MLCCs can be used in PCBs rated to a voltage range of a few volts to several hundred volts, depending on the application.

1.5.1.3 Mica Capacitor

Mica capacitors are capacitors that use mica, a group of natural minerals with very high electrical, chemical, and mechanical stability as the dielectric. The typical layered structure of mica allows the manufacture of thin sheets in the order of 0.025–0.125 mm. The capacitance value of these capacitors lies between pF to nF. There are two types of mica capacitors available on the market, namely clamped mica capacitors and silver mica capacitors. Clamped mica capacitors are not widely used nowadays due to their inferior characteristics, however, silver mica capacitors are used instead. The construction of these capacitors involves sandwiching mica sheets with metal on both sides. Then this assembly is enclosed in epoxy to protect it from the harsh environment. The advantages of mica capacitors include very high stability and accuracy, frequency-independent characteristics, and low resistive and inductive losses. Silver mica capacitors are used in applications that require low capacitance values with high stability and low losses such as in power radio frequency circuits and also in high frequency tuned circuits, like filters and oscillators. Also, these capacitors are bulky and expensive.

1.5.1.4 Paper/Plastic Film Capacitor

A paper capacitor, otherwise known as fixed capacitor, uses paper as a dielectric medium. It stores a fixed amount of electric charge with a capacitance value of 1 nF to 1 μF. The construction of a paper capacitor is similar to a mica capacitor. Here, paper sheets and two aluminum sheets rolled in a cylindrical shape constitute the cell assembly and it is then coated with wax/plastic resin. Then two-wire leads are taken out from the ends of the aluminum sheets. In a metalized paper capacitor, the paper is coated with a thin layer of zinc or aluminum and rolled in the form of a cylinder. Disadvantages of paper capacitors include variation in capacitance with temperature and short service life.

A plastic film capacitor uses some plastic materials such as Teflon, polystyrene, polyethylene terephthalate, and polypropylene as dielectrics and metals like zinc or aluminum as the electrodes of the capacitor. The two types of plastic film capacitors are film-foil capacitors and metalized film capacitors. The construction of the film-foil capacitor requires two plates of aluminum foil separated by insulating plastic films as the dielectric. Whereas in metalized film capacitors aluminum or zinc is directly coated onto the dielectric using a vacuum deposition process. These types of capacitors are associated with self-healing property which prevents dielectric breakdown by a process that removes defects in the dielectric film by vaporizing the metal electrode surrounding that defect [26]. The wide applicability of film capacitors could be attributed to their high stability, low equivalent series resistance (ESR), low self-inductance, and low cost. These film capacitors provide a precise value of capacitance in the range of nF to mF, have a long life and high reliability, and also operate under high-temperature conditions. The capacitance versus voltage graph of the different classes of capacitors is provided in Figure 1.1(d).

1.5.2 Electrolytic Capacitor

The history of electrolytic capacitors started in 1886 when Charles Pollak anodized metals like aluminum and other metals (Figure 1.1(b)). During his research, he observed that there is very high capacitance between the aluminum and the electrolyte solution due to the thin layer of aluminum oxide produced during anodization. An electrolytic capacitor is a type of polarized capacitor whose positive electrode is made of a metal capable of forming an insulating layer of oxide during anodization and uses an electrolyte as the negative electrode to achieve a greater capacitance than other types of capacitors. Here the oxide layer constitutes the dielectric of the capacitor.

The electrolyte can be a liquid, semi-solid, or solid with a high concentration of ions. The polarity of most electrolytic capacitors arises because the voltage on the anode on which the dielectric layer formed must always be greater than the voltage on the cathode. The process of anodization sets the polarity. So, care must be taken over the polarity of the capacitor in a circuit. If the reversal of polarities occurs, it triggers dissolution of the oxide layer, and in extreme cases the device may explode owing to overheating of the electrolyte. Different types of electrolytic capacitors are available on the market.

1.5.2.1 Aluminum Electrolytic Capacitor

An aluminum electrolytic capacitor consists of two aluminum foils as electrodes and electrolytes like a solution of phosphate, carbonate, or borax. When a DC voltage is passed across the plates of the capacitor, the metal oxide is of the order of a micrometer formed by the electrochemical oxidation of the aluminum foil which acts as the dielectric of the capacitor. These types of capacitors are found in many applications like computer motherboards and power supplies. Based on the working voltage, applications, and type of electrolyte used, these capacitors are available in different sizes and shapes. Their capacitance values lie between 0.1 μF and 2.7 F with voltage rated from 4 to 630 V. The electrolytic solutions for aluminum capacitors must have properties such as the high ability to self-repair the oxide film on aluminum, high electro conduction, and a broad range of temperature stability [27]. The electrolytes for aluminum capacitors vary extensively. These include salts of adipic, trifluoroacetic, tartaric, maleic, salicylic, citric, and formic acids [28]. Organic electrolytes based on gamma-butyrolactone, ethylene carbonate, and other compounds are also used in aluminum capacitors [27, 29]. Solid electrolytes like salts of 7,7,8,8-tetracyanoquinonodimethane complex [30–33], N-n-butyl isoquinolin [34], and solid manganese oxide are also used. Even though electrolytic capacitors have large capacitance compared with other capacitors, they have several drawbacks as well. The major disadvantages of aluminum electrolytic capacitors include limited lifetime, the dependence of their parameters on temperature, and considerable leakage current, value tolerances, and equivalent series resistance.

1.5.2.2 Tantalum and Niobium Electrolytic Capacitor

A new version of an electrolytic capacitor has emerged, with tantalum metal having a superior property in which tantalum metal forms the anode of the capacitor covered by a very thin layer of tantalum pentoxide as the dielectric and surrounded by a conductive liquid or solid electrolyte as the cathode. The main features of these capacitors compared to aluminum capacitors are higher specific capacitance, low equivalent series resistance, reduction of ignition risk, self-healing reaction, smaller leakage currents, better temperature–frequency characteristics, and longer storage times. Conventional tantalum capacitors with a solid electrolyte like MnO_2 are considered to be the ideal choice for high temperature applications (currently up to 175°C) and also when high voltage is required (up to 50 V). On the other hand, tantalum polymer capacitors are used in consumer applications such as telecommunications, personal digital assistants, etc., wherein ESR requirements are quite low. The significantly larger mechanical strength of tantalum allows the construction of a thinner foil, which additionally increases the capacitance. However, tantalum capacitors are more expensive than aluminum ones [35, 36].

At present, work is in progress to check the feasibility of the oxides of niobium and titanium, which possess a high dielectric constant and have a lower cost than tantalum oxide, in electrolytic capacitor applications. The construction of niobium electrolytic capacitors is similar to tantalum ones but here niobium pentoxide acts as the dielectric and niobium monoxide or niobium metal forms the anode. NbO capacitors are characterized by good cost-versus-performance value, also they have the advantages of lower cost and the better availability of niobium compared to tantalum. The excellent reliability of these capacitors makes them a favorite choice for automotive and computer applications.

1.5.3 SUPERCAPACITORS

Currently, technological advancement in many fields requires the need for storage devices with improved storage capacity to store energy when it is available and regain when it is required. This situation has forced humanity to develop a new energy storage system that couples higher energy density than existing dielectric capacitors and higher power density than batteries known as supercapacitors [37, 38]. The history of supercapacitors began with the patent describing double-layer capacitance in 1957. Since these electrochemical capacitors use electrolyte solutions and high-surface-area electrodes, they have higher capacitance per unit volume than conventional dielectric capacitors and electrolytic capacitors. The very high surface area associated with electrodes and the small separation of the order of atom or molecule between the electronic and ionic charge at the electrode surface allow supercapacitors to have higher energy density than conventional counterparts [39, 40]. The Ragone plot of various electrochemical systems shows that fuel cells are associated with high energy density and supercapacitors with high power density, whereas batteries lie in between them with intermediate energy and power density [41, 42]. They are highly recommended for applications that require energy pulses for a short period like automobiles, cranes, forklifts, electricity load-leveling in stationary and transportation systems, etc. [43, 44]. The other name for supercapacitors is ultracapacitors, which have a higher capacitance value than the other types of capacitors [45]. The major components of a supercapacitor include an electrolyte, two electrodes, and a separator which isolates (electrically) the two electrodes. Even though supercapacitors are associated with advantages like higher power density and better lifecycle, they have unsatisfactory low energy density [46, 47], a low amount of energy stored per unit weight, higher dielectric absorption, and a higher rate of self-discharge than that an electrochemical battery. Also, its very low internal resistance causes extremely rapid discharge when shorted, resulting in a spark hazard. In addition, these supercapacitors are also prone to the process of degradation and corrosion especially at high charge voltage. Based on the mechanism of energy storage and the type of electrode material, supercapacitors are divided into three groups, namely electric double-layer capacitors (EDLCs), pseudocapacitors, and hybrid capacitors [48–50].

1.5.3.1 Electric Double-Layer Capacitor (EDLC)

The main components of EDLCs (Figure 1.1(c)) consist of two porous electrodes, an electrolyte, and a separator which electrically insulate the positive and negative electrodes [51]. In EDLCs, as the name indicates, electrical double-layer capacitance originates from the separation of electric charges, caused by the directional distribution of ions and electrons at the interface between the electrode material and electrolyte [52]. The double-layer shows two regions of ion distributions, namely the compact Stern layer, or the inner Helmholtz plane, and the diffuse layer or outer Helmholtz plane. Surface dissociation and ion adsorption of the electrolyte and crystal lattice defects contribute to the surface electrode charge generation. During the charging process, the positively charged ions in the electrolyte migrate towards the negative electrode and the negatively charged ions migrate to the positive electrode, while the electrons are transferred from the negative electrode to the positive electrode through an external current source [53]. When the external load has removed the stabilization of the electrical double layer on the electrodes, it occurs by the attraction of oppositely charged ions in the electrolyte and the charges on the electrode surface. The reverse process occurs during the discharging process. The net result of this process is the accumulation of energy in the double-layer interface. Since there is no electrochemical reaction between electrode materials and only physical charge accumulation takes place at the solid/electrolyte interface during the charging/discharging processes, EDLCs can sustain the very large number of cycles of the order of millions with high energy density. In view of this, electric double-layer charge storage can be considered as a surface phenomenon. So, the capacitance of these types of capacitors can be affected by the surface properties of the electrode materials. Hence, according to this mechanism the electrode material plays an important role in the performance of EDLCs. Therefore, it is very important to select the appropriate electrode material for EDLCs. Different forms of carbon material such as powder, fibers, porous carbon, carbon nanotubes (CNTs), graphene, etc., are widely used as electrode material in EDLCs. The easy availability and existence in various allotropic forms made carbon material more attractive for capacitor applications [54–64].

1.5.3.2 Pseudocapacitor

The way of storing charge in a pseudocapacitor or faradaic supercapacitor is different from EDLCs [65]. Here the capacitance arises from fast and reversible faradaic reactions or redox reactions at or near the electrode surface [66–68]. Three types of electrochemical processes occur for the development of pseudocapacitance. These include reversible surface adsorption/desorption of protons or any metal ions from the electrolyte, redox reactions involving a charge transfer with the electrolyte, and reversible doping/undoping of an active conducting polymer material of the electrode. The surface area of the electrodes plays important role in the first two processes because they are primarily surface reactions. The third process involving the conducting polymer is a bulk process, it does not depend on the surface area of the electrodes; however, a material with a suitable pore structure, preferably a micropore structure, is desirable for the to-and-fro movement of ions from the electrodes in a cell. Compared with a battery the charging–discharging of pseudocapacitive material happens quickly, normally within seconds or minutes. Batteries will take a much longer time for the charging and discharging process. Hence pseudocapacitive material can be used as high energy and power density material. Initially, the focus of research in this area was based on electrosorption of proton monolayers on noble metals like platinum or gold and electrochemical protonation of metal hydrous oxides such as iridium and ruthenium oxide hydrates [69–70]. Recently, the focus changed to the materials undergoing such redox reactions, including conducting polymers and several transition metal oxides, including RuO_2 [71], MnO_2 [72–74], V_2O_5 [75–77], and Co_3O_4, and conducting polymers like, e.g., polyaniline, polypyrrole, polyvinyl alcohol, etc. [78–80]. In principle, systems with multiple oxidation state pseudocapacitors can provide a higher energy density than EDLCs. But the pseudocapacitors have relatively poor durability compared with EDLCs owing to the physical changes that occur during the charge/discharge cycle.

1.5.3.3   Hybrid Supercapacitor

The pseudocapacitance can be suitably coupled on any EDLC, which forms the basis of the third type of supercapacitor known as a hybrid capacitor, which recently come into the picture and progressed rapidly [81–90]. Typical hybrid capacitors consist of a cathode and anode composed of different active materials [91], a separator which isolates the two electrodes, and an electrolyte. Hybrid supercapacitors consisting of a redox (faradaic) reaction electrode or a battery-type electrode, with an electric double-layer electrode (typically carbon materials) in suitable electrolyte are employed to further improve the energy density [91–93]. Here the charge-storing mechanism involves both double-layer ion adsorption/desorption and reversible faradaic reaction. Such asymmetric configurations have the redox-type electrodes with a large energy density, while the non-faradaic capacitive electrode has high power density and excellent cycling stability [94]. Hybrid capacitors were divided into three categories differentiated via their electrode configurations as asymmetric, composite, and battery type.

1.6   SUMMARY AND OUTLOOK

Nowadays the worldwide research community is working hard to improv the properties of energy storage systems with the help of nanotechnology to enhance the capacity of storing energy from renewable sources of energy like solar and wind. The fast-growing consumer electronics which demand thinner, lighter, and flexible storage units are not satisfied with the rigidity and large size of traditional capacitors. Supercapacitors and batteries are presently handling these requirements to a certain extent. Even though much research has taken place in the field of supercapacitors, it has to go a lot further for significant energy density. The focus of research in this area is to amplify the characteristics of all types of supercapacitors and batteries in terms of cycle stability, specific energy, specific power, and to reduce production costs. Currently, the world of energy storage is eagerly looking for various hybrid energy storage systems which combine two or more types of energy storage systems like batteries and capacitors. At the present stage, a single energy storage system cannot fulfill all the desired operations due to its limited capability and potency in terms of lifespan, cost, energy, and power density, etc. Hence, different configurations of hybrid systems, especially battery and supercapacitor systems, can use the advantages of each system to improve overall performance to meet future needs.

REFERENCES

1. Gilliam, R. (2002). Revisiting the winning of the west. *Bulletin of Science, Technology & Society*, 22(2), 147–157.
2. Winter, M., & Brodd, R. J. (2004). What are batteries, fuel cells, and supercapacitors?. *Chemical Reviews*, 104(10), 4245–4270
3. Arunachalam, V. S., & Fleischer, E. L. (2008). Harnessing materials for energy. Chemistry International--Newsmagazine for IUPAC, 30(4), 25–26.
4. Armand, M., & Tarascon, J. M. (2008). Building better batteries. *Nature*, 451(7179), 652–657.
5. Tollefson, J. (2008). Car industry: Charging up the future. *Nature News*, 456(7221), 436–440.
6. Jacobson, M. Z. (2009). Review of solutions to global warming, air pollution, and energy security. *Energy & Environmental Science*, 2(2), 148–173.
7. Tan, Q., Irwin, P., & Cao, Y. (2006). Advanced dielectrics for capacitors. *IEEJ Transactions on Fundamentals and Materials*, 126(11), 1153–1159.
8. Arico, A. S., Bruce, P., Scrosati, B., Tarascon, J. M., & Van Schalkwijk, W. (2011). Nanostructured materials for advanced energy conversion and storage devices. Materials for sustainable energy: *Nature Materials*, 148–159.
9. Jiang, C., Hosono, E., & Zhou, H. (2006). Nanomaterials for lithium ion batteries. *Nano Today*, 1(4), 28–33.
10. Bruce, P. G., Scrosati, B., & Tarascon, J. M. (2008). Nanomaterials for rechargeable lithium batteries. *AngewandteChemie International Edition*, 47(16), 2930–2946.
11. Guo, Y. G., Hu, J. S., & Wan, L. J. (2008). Nanostructured materials for electrochemical energy conversion and storage devices. *Advanced Materials*, 20(15), 2878–2887.
12. Simon, P., & Gogotsi, Y. (2010). Materials for electrochemical capacitors. *Nature Materials*, 320–329.
13. Rolison, D. R., Long, J. W., Lytle, J. C., Fischer, A. E., Rhodes, C. P., McEvoy, T. M., Megan E. B., & Lubers, A. M. (2009). Multifunctional 3D nanoarchitectures for energy storage and conversion. *Chemical Society Reviews*, 38(1), 226–252.
14. Wang, Y., Li, H., He, P., Hosono, E., & Zhou, H. (2010). Nano active materials for lithium-ion batteries. *Nanoscale*, 2(8), 1294–1305.
15. Burke, A. (2000). Ultracapacitors: Why, how, and where is the technology. *Journal of Power Sources*, 91(1), 37–50.
16. David Resnick, Halliday, Robert, & Walker, Jearl. (2020). *Halliday and Resnick's Principles of Physics*. John Wiley & Sons.
17. Diagram comparing construction of three types of capacitors: Electrostatic (normal), electrolytic (high capacity) and electrochemical (supercapacitors), Maxwell Technologies, Wikipedia (2006). https://commons.wikimedia.org/ wiki/File: Superca pacitordiagram.svg#filelinks
18. Elcap, Capacitance ranges vs voltage ranges of different capacitor types, in: *Wikipedia*, (2012), https://commons.wikimedia.org/wiki/File:Kondensatoren-Kap-Versus-Spg -English.svg.

19. Choi, J. M., & Kim, T. W. (2013). Humidity sensor using an air capacitor. *Transactions on Electrical and Electronic Materials*, 14(4), 182–186.

20. Pan, M. J., & Randall, C. A. (2010). A brief introduction to ceramic capacitors. *IEEE Electrical Insulation Magazine*, 26(3), 44–50.

21. Cheng, C. M., Lo, S. H., & Yang, C. F. (2000). The effect of CuO on the sintering and properties of BiNbO4 microwave ceramics. *Ceramics International*, 26(1), 113–117.

22. Burn, I., & Smyth, D. M. (1972). Energy storage in ceramic dielectrics. *Journal of Materials Science*, 7(3), 339–343.

23. Li, W., Auciello, O., Premnath, R. N., & Kabius, B. (2010). Giant dielectric constant dominated by Maxwell–Wagner relaxation in Al $_2O_3$/TiO$_2$nanolaminates synthesized by atomic layer deposition. *Applied physics letters*, 96(16), 162907.

24. Wang, X., Zhang, Y., Song, X., Yuan, Z., Ma, T., Zhang, Q., Deng, C. & Liang, T. (2012). Glass additive in barium titanate ceramics and its influence on electrical breakdown strength in relation with energy storage properties. *Journal of the European Ceramic Society*, 32(3), 559–567.

25. Iguchi, Y., Honda, M., & Kishi, H. (1999). U.S. Patent No. 5,977,006. Washington, DC: U.S. Patent and Trademark Office.

26. Tortai, J. H., Denat, A., & Bonifaci, N. (2001). Self-healing of capacitors with metallized film technology: Experimental observations and theoretical model. *Journal of Electrostatics*, 53(2), 159–169.

27. Monta, M., & Matsuda, Y. (1996). Ethylene carbonate-based organic electrolytes for high performance aluminium electrolytic capacitors. *Journal of power sources*, 60(2), 179–183.

28. Bernard, W. J. (1977). Developments in electrolytic capacitors. *Journal of the Electrochemical Society*, 124(12), 403C.

29. Ue, M., Takeda, M., Suzuki, Y., & Mori, S. (1996). Chemical stability of γ-butyrolactone-based electrolytes for aluminum electrolytic capacitors. *Journal of Power Sources*, 60(2), 185–190.

30. Kyokane, J., & Yoshino, K. (1996). Organic solid capacitor with conducting thin films as electrolyte by ion-beam-assisted deposition. *Journal of Power Sources*, 60(2), 151–155.

31. Kyokane, J., Taniguchi, I., & Yoshino, K. (1995). Conducting organic thin films by ion-assisted evaporation and their application to electronic devices and components. *Synthetic Metals*, 71(1–3), 2219–2220.

32. Kyokane, J., Aoyagi, R., & Yoshino, K. (1997). Application to electronic devices using organic thin films by ion-beam-evaporation method. *Synthetic Metals*, 85(1–3), 1393–1394.

33. Vol'fkovich, Y. M., & Serdyuk, T. M. (2002). Electrochemical capacitors. *Russian Journal of Electrochemistry*, 38(9), 935–959.

34. Kudoh, Y., Tsuchiya, S., Kojima, T., Fukuyama, M., & Yoshimura, S. (1991). An aluminum solid electrolytic capacitor with an electroconducting-polymer electrolyte. *Synthetic Metals*, 41(3), 1133–1136

35. Martinez-Duart, J. M., Velilla, J. L., Albella, J. M., & Rueda, F. (1974). Dielectric properties of thin Ta$_2$O$_5$ films. *Physica Status Solidi (A)*, 26(2), 611–615.

36. Goudswaard, B., & Driesens, F. J. J. (1976). Failure mechanism of solid tantalum capacitors. *Electrocomponent Science and Technology*, 3(3), 171–179.

37. Conway, B. E. (2013). *Electrochemical Supercapacitors: Scientific Fundamentals and Technological Applications*. Springer Science & Business Media.

38. Conway, B. E. (1991). Transition from "supercapacitor" to "battery" behavior in electrochemical energy storage. *Journal of the Electrochemical Society*, 138(6), 1539.

39. Kötz, R., & Carlen, M. J. E. A. (2000). Principles and applications of electrochemical capacitors. *Electrochimicaacta*, 45(15–16), 2483–2498.

40. Bard, A. J., & Faulkner, L. R. (2001). Fundamentals and applications. *Electrochemical Methods*, 2(482), 580–632.
41. Winter, M., & Brodd, R. J. (2004). What are batteries, fuel cells, and supercapacitors?. *Chemical Reviews*, 104(10), 4245–4270.
42. Christen, T., & Carlen, M. W. (2000). Theory of Ragone plots. *Journal of Power Sources*, 91(2), 210–216.
43. Stoller, M. D., Park, S., Zhu, Y., An, J., & Ruoff, R. S. (2008). Graphene-based ultracapacitors. *Nano Letters*, 8(10), 3498–3502.
44. Miller, J. R., & Simon, P. (2008). Electrochemical capacitors for energy management. *Science Magazine*, 321(5889), 651–652.
45. Pandolfo, A. G., & Hollenkamp, A. F. (2006). Carbon properties and their role in supercapacitors. *Journal of Power Sources*, 157(1), 11–27.
46. Qu, Q. T., Shi, Y., Tian, S., Chen, Y. H., Wu, Y. P., & Holze, R. (2009). A new cheap asymmetric aqueous supercapacitor: Activated carbon//NaMnO2. *Journal of Power Sources*, 194(2), 1222–1225.
47. Qu, Q., Li, L., Tian, S., Guo, W., Wu, Y., & Holze, R. (2010). A cheap asymmetric supercapacitor with high energy at high power: Activated carbon//K0. 27MnO2· 0.6 H2O. *Journal of Power Sources*, 195(9), 2789–2794.
48. Conway, B. E. (1996). Supercapacitorbehavior resulting from pseudocapacitance associated with redox processes. In *Proceedings of the Symposium on Electrochemical Capacitors* (Vol. 95, p. 29).
49. Burke, A. (2000). Ultracapacitors: Why, how, and where is the technology. *Journal of Power Sources*, 91(1), 37–50.
50. Zhong, C., Deng, Y., Hu, W., Qiao, J., Zhang, L., & Zhang, J. (2015). A review of electrolyte materials and compositions for electrochemical supercapacitors. *Chemical Society Reviews*, 44(21), 7484–7539.
51. Frackowiak, E., & Beguin, F. (2001). Carbon materials for the electrochemical storage of energy in capacitors. *Carbon*, 39(6), 937–950.
52. Zhang, L. L., & Zhao, X. S. (2009). Carbon-based materials as supercapacitor electrodes. *Chemical Society Reviews*, 38(9), 2520–2531.
53. Wang, G., Zhang, L., & Zhang, J. (2012). A review of electrode materials for electrochemical supercapacitors. *Chemical Society Reviews*, 41(2), 797–828.
54. Futaba, D. N., Hata, K., Yamada, T., Hiraoka, T., Hayamizu, Y., Kakudate, Y., Tanaike, O., Hatori, H., Yumura, T., & Iijima, S. (2006). Shape-engineerable and highly densely packed single-walled carbon nanotubes and their application as super-capacitor electrodes. *Nature Materials*, 5(12), 987–994.
55. Lee, J., Sohn, K., & Hyeon, T. (2001). Fabrication of novel mesocellular carbon foams with uniform ultralarge mesopores. *Journal of the American Chemical Society*, 123(21), 5146–5147.
56. Cheng, H., Dong, Z., Hu, C., Zhao, Y., Hu, Y., Qu, L, Nan, C. & Dai, L. (2013). Textile electrodes woven by carbon nanotube–graphene hybrid fibers for flexible electrochemical capacitors. *Nanoscale*, 5(8), 3428–3434.
57. Okajima, K., Ohta, K., & Sudoh, M. (2005). Capacitance behavior of activated carbon fibers with oxygen-plasma treatment. *ElectrochimicaActa*, 50(11), 2227–2231.
58. Chen, L. F., Zhang, X. D., Liang, H. W., Kong, M., Guan, Q. F., Chen, P., Zhen-Yu Wu. & Yu, S. H. (2012). Synthesis of nitrogen-doped porous carbon nanofibers as an efficient electrode material for supercapacitors. *ACS Nano*, 6(8), 7092–7102.
59. Wang, D. W., Li, F., Liu, M., Lu, G. Q., & Cheng, H. M. (2008). 3D aperiodic hierarchical porous graphitic carbon material for high-rate electrochemical capacitive energy storage. *AngewandteChemie International Edition*, 47(2), 373–376.

60. Niu, C., Sichel, E. K., Hoch, R., Moy, D., & Tennent, H. (1997). High power electrochemical capacitors based on carbon nanotube electrodes. *Applied Physics Letters*, 70(11), 1480–1482.

61. Kaempgen, M., Chan, C. K., Ma, J., Cui, Y., & Gruner, G. (2009). Printable thin film supercapacitors using single-walled carbon nanotubes. *Nano Letters*, 9(5), 1872–1876.

62. Liu, J. (2014). *Charging Graphene for Energy Storage (No. PNNL-SA-105402)*. Pacific Northwest National Lab. (PNNL), Richland, WA (United States).

63. El-Kady, M. F., Strong, V., Dubin, S., & Kaner, R. B. (2012). Laser scribing of high-performance and flexible graphene-based electrochemical capacitors. *Science*, 335(6074), 1326–1330.

64. Miller, J. R., Outlaw, R. A., & Holloway, B. C. (2010). Graphene double-layer capacitor with ac line-filtering performance. *Science*, 329(5999), 1637–1639.

65. Wang, H., Casalongue, H. S., Liang, Y., & Dai, H. (2010). Ni (OH) 2 nanoplates grown on graphene as advanced electrochemical pseudocapacitor materials. *Journal of the American Chemical Society*, 132(21), 7472–7477.

66. Zhong, C., Deng, Y., Hu, W., Qiao, J., Zhang, L., & Zhang, J. (2015). A review of electrolyte materials and compositions for electrochemical supercapacitors. *Chemical Society Reviews*, 44(21), 7484–7539.

67. Burke, A. (2000). Ultracapacitors: Why, how, and where is the technology. *Journal of Power Sources*, 91(1), 37–50.

68. Brousse, T., Bélanger, D., & Long, J. W. (2015). To be or not to be pseudocapacitive. *Journal of the Electrochemical Society*, 162(5), A5185.

69. Hadz, S., Angerstein-Kozlowska, H., Vukovič, M., & Conway, B. E. (1978). Reversibility and growth behavior of surface oxide films at ruthenium electrodes. *Journal of the Electrochemical Society*, 125(9), 1471.

70. Mozota, J., & Conway, B. E. (1983). Surface and bulk processes at oxidized iridium electrodes—I. Monolayer stage and transition to reversible multilayer oxide film behaviour. *Electrochimica Acta*, 28(1), 1–8.

71. Trasatti, S., & Buzzanca, G. (1971). Ruthenium dioxide: A new interesting electrode material. Solid state structure and electrochemical behaviour. *Journal of Electroanalytical Chemistry and Interfacial Electrochemistry*, 29(2), A1–A5.

72. Biswal, A., Tripathy, B. C., Sanjay, K., Subbaiah, T., & Minakshi, M. (2015). Electrolytic manganese dioxide (EMD): A perspective on worldwide production, reserves and its role in electrochemistry. *RSC Advances*, 5(72), 58255–58283.

73. Ma, S. B., Nam, K. W., Yoon, W. S., Yang, X. Q., Ahn, K. Y., Oh, K. H., & Kim, K. B. (2008). Electrochemical properties of manganese oxide coated onto carbon nanotubes for energy-storage applications. *Journal of Power Sources*, 178(1), 483–489.

74. Qiu, Y., Xu, P., Guo, B., Cheng, Z., Fan, H., Yang, M, Xiaoxi Yang & Li, J. (2014). Electrodeposition of manganese dioxide film on activated carbon paper and its application in supercapacitors with high rate capability. *RSC Advances*, 4(109), 64187–64192.

75. Dong, W., Rolison, D. R., & Dunna, B. (2000). Electrochemical properties of high surface area vanadium oxide aerogels. *Electrochemical and Solid State Letters*, 3(10), 457.

76. Nakamura, A. (2010). New defect-crystal-chemical approach to non-Vegardianity and complex defect structure of fluorite-based MO2–LnO1. 5 solid solutions (M4+= Ce, Th; Ln3+= lanthanide): Part II: Detailed local-structure and ionic-conductivity analysis. *Solid State Ionics*, 181(37–38), 1631–1653.

77. Kudo, T., Ikeda, Y., Watanabe, T., Hibino, M., Miyayama, M., Abe, H., & Kajita, K. (2002). Amorphous V2O5/carbon composites as electrochemical supercapacitor electrodes. *Solid State Ionics*, 152, 833–841.

78. Wu, M. S., & Chiang, P. C. J. (2004). Fabrication of nanostructured manganese oxide electrodes for electrochemical capacitors. *Electrochemical and Solid State Letters*, 7(6), A123.

79. Sugimoto, W., Iwata, H., Murakami, Y., & Takasu, Y. (2004). Electrochemical capacitor behavior of layered ruthenic acid hydrate. *Journal of The Electrochemical Society*, 151(8), A1181.

80. Dong, X., Shen, W., Gu, J., Xiong, L., Zhu, Y., Li, H., & Shi, J. (2006). MnO2-embedded-in-mesoporous-carbon-wall structure for use as electrochemical capacitors. *The Journal of Physical Chemistry B*, 110(12), 6015–6019.

81. Wang, X., Kajiyama, S., Iinuma, H., Hosono, E., Oro, S., Moriguchi, I ;, Masashi Okubo & Yamada, A. (2015). Pseudocapacitance of MXenenanosheets for high-power sodium-ion hybrid capacitors. *Nature Communications*, 6(1), 1–6.

82. Yang, J. J., Choi, C. H., Seo, H. B., Kim, H. J., & Park, S. G. (2012). Voltage characteristics and capacitance balancing for Li4Ti5O12/activated carbon hybrid capacitors. *ElectrochimicaActa*, 86, 277–281.

83. Conway, B. E., & Pell, W. G. (2003). Double-layer and pseudocapacitance types of electrochemical capacitors and their applications to the development of hybrid devices. *Journal of Solid State Electrochemistry*, 7(9), 637–644.

84. Kim, J. H., Kang, S. H., Zhu, K., Kim, J. Y., Neale, N. R., & Frank, A. J. (2011). Ni–NiO core–shell inverse opal electrodes for supercapacitors. *Chemical Communications*, 47(18), 5214–5216

85. Jiang, H., Yang, L., Li, C., Yan, C., Lee, P. S., & Ma, J. (2011). High–rate electrochemical capacitors from highly graphitic carbon–tipped manganese oxide/mesoporous carbon/manganese oxide hybrid nanowires. *Energy & Environmental Science*, 4(5), 1813–1819.

86. Jiang, J., Tan, G., Peng, S., Qian, D., Liu, J., Luo, D., & Liu, Y. (2013). Electrochemical performance of carbon-coated Li3V2 (PO4) 3 as a cathode material for asymmetric hybrid capacitors. *ElectrochimicaActa*, 107, 59–65.

87. Lu, K., Song, B., Gao, X., Dai, H., Zhang, J., & Ma, H. (2016). High-energy cobalt hexacyano ferrate and carbon micro-spheres aqueous sodium-ion capacitors. *Journal of Power Sources*, 303, 347–353

88. Khomenko, V., Raymundo-Piñero, E., & Béguin, F. (2008). High-energy density graphite/AC capacitor in organic electrolyte. *Journal of Power Sources*, 177(2), 643–651.

89. Chen, L. M., Lai, Q. Y., Hao, Y. J., Zhao, Y., & Ji, X. Y. (2009). Investigations on capacitive properties of the AC/V2O5 hybrid supercapacitor in various aqueous electrolytes. *Journal of Alloys and Compounds*, 467(1–2), 465–471.

90. Wang, X., Li, M., Chang, Z., Yang, Y., Wu, Y., & Liu, X. (2015). Co3O4@ MWCNT nanocable as cathode with superior electrochemical performance for supercapacitors. *ACS Applied Materials & Interfaces*, 7(4), 2280–2285.

91. Wang, F., Xiao, S., Hou, Y., Hu, C., Liu, L., & Wu, Y. (2013). Electrode materials for aqueous asymmetric supercapacitors. *RSC Advances*, 3(32), 13059–13084.

92. Wang, F., Wang, X., Chang, Z., Wu, X., Liu, X., Fu, L., Wu, Y., & Huang, W. (2015). A quasi-solid-state sodium-ion capacitor with high energy density. *Advanced Materials*, 27(43), 6962–6968.

93. Hong, M. S., Lee, S. H., & Kim, S. W. (2002). Use of KCl aqueous electrolyte for 2 V manganese oxide/activated carbon hybrid capacitor. *Electrochemical and Solid State Letters*, 5(10), A227.

94. Aravindan, V., Gnanaraj, J., Lee, Y. S., & Madhavi, S. (2014). Insertion-type electrodes for nonaqueous Li-ion capacitors. *Chemical Reviews*, 114(23), 11619–11635.

2 Electric Double-Layer Capacitors

*Shyamli Ashok, Deepu Joseph, Jasmine Thomas,
Nygil Thomas and Anjali Paravannoor*

CONTENTS

2.1 INTRODUCTION

EDLC/goldcaps/ultracapacitors/supercapacitors have been considered to be a well-tried alternative energy storage device for the last few decades. Owing to their power profile and ultra-long lifespan, the demand for novel designs and configurations in electric double-layer capacitors (EDLCs) is rapidly increasing and is especially so as there are diverse application areas being developed, constantly expanding the market [1–4]. Their function also varies depending upon the specific application, wherein it would augment the battery technology or replace it.

DOI: 10.1201/9781003258384-2

The energy storage in EDLCs could be attributed to the charge accumulation in the electrical double layer (EDL) at the electrode/electrolyte interface wherein the separation between the electronic charge at the electrode surface and the ionic charge at the electrolyte region causes electrostatic charge separation [5]. However, compared to the conventional parallel plate capacitor which uses a dielectric material, EDLCs form an insulating layer of the solvent, localizing the electric field at the angstrom level and macroscopically it assumes the same equation [6]:

$$C = \frac{\varepsilon_0 \varepsilon_r}{d} A \qquad (2.1)$$

so that a high surface area electrode would provide an extraordinary capacitance value. At the same time, this electrostatic mechanism would allow the device a quick response which in turn would improve the power profile with a power density as high as 15 kW/kg on average and a very long lifespan in the order of 1,000,000 cycles.

2.2   ELECTRICAL DOUBLE-LAYER THEORIES

When a charged object (electrode) is in contact with a liquid (electrolyte), a layer of opposite charges from the liquid concentrates near the surface of the object to balance the charge on it, to form an electrical double layer at the interface separating the electrode and the electrolyte. The formation of such a layer at the interface and the subsequent interaction between the electrode surface and electrolyte ions are explained by many different models and theories. Three prominent models are given below.

2.2.1   THE HELMHOLTZ MODEL

Hermann von Helmholtz proposed that when an electrode is in contact with an electrolyte, ions of opposite charge tend to attract while ions of same charge repel.

This results in the formation of two layers at the electrode/electrolyte interface and is called an electrical double layer. This electrical double layer can be considered as a molecular dielectric that can store an electric charge. The value of capacity for this molecular dielectric can be written as:

$$C = \frac{\varepsilon_0 \varepsilon_r}{X_H} A \qquad (2.2)$$

where X_H is the ionic radius (distance of closest approach of the charges), ε_r is the relative permittivity (dielectric constant), ε_0 is the permittivity of vacuum and A is the surface area accessible to the electrolyte ions. The value of capacity does not vary with applied potential the potential falls linearly from φ_M to φ_S where φ_M is the value of electrostatic potential in the metal electrode and φ_S its value in the solution (Figure 2.1(a)). This is the simplest approach to interpret a double layer. But this model does not consider diffusion of ions, adsorption of ions at the surface of the electrode and the interaction between dipole moments of the electrode and the solvent [7, 8].

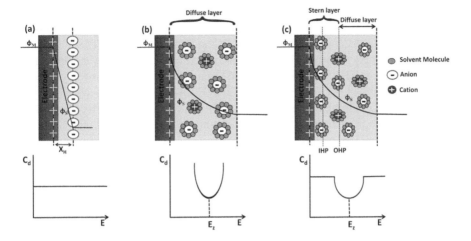

FIGURE 2.1 The Helmholtz (a), Gouy-Chapman (b) and Stern (c) models of the double-layer variation of C_d with potential are shown.

2.2.2 THE GOUY-CHAPMAN OR DIFFUSE MODEL

According to Louis Georges Gouy and David Leonard Chapman, the value of capacitance depends on both ionic concentration and applied potential. As per this model, the charged surface is surrounded by equal numbers of oppositely charged ions from solution, but these ions are not bound to the charged surface as described in the Helmholtz model, instead, they are free to move. This layer of oppositely charged ionsis therefore called a diffuse double layer. The kinetic energy of the ions in the solution is one of the critical factors that determines the thickness of this layer. The distribution of ionic species with distance from the electrode follows Maxwell–Boltzmann's law:

$$n_i = n_i^0 \exp\left(\frac{Z_i e \varphi \Delta}{K_B T}\right) \tag{2.3}$$

where n_i^0 is the numerical concentration of ions i in bulk solution and $\varphi_\Delta = \varphi - \varphi_S$. Therefore, electric potential falls exponentially from surface to the bulk of the solution as shown in Figure 2.1(b) [7, 8]. However, this model fails to explain the highly charged double layer where the calculated value of double-layer thickness is less than the experimental value.

2.2.3 THE STERN MODEL

This model combines the features of the Helmholtz model and the Gouy–Chapman model. According to the Stern model, some ions are attached rigidly to the surface of the electrode like those in the Helmholtz model, which is called a Stern layer, whereas some others form the diffuse layer as proposed by Gouy and Chapman. The Gouy–Chapman model considers ions as point charges and there is no restriction for them to approach the surface of the electrode. Stern modified this assumption by considering the finite size of the ions [9]. The Stern model depicts that the first layer of ions (Stern layer) are at a distance δ from the electrode surface. The Stern layer consists of strongly adsorbed electrolyte ions which form a narrow layer called inner Helmholtz plane (IHP). A weakly adsorbed layer adjacent to this narrow layer forms the outer Helmholtz plane (OHP). The effective capacity (C_d) of the system can be determined by assuming the series combination of two capacitors with capacities C_H representing the Stern layer and C_{GC} representing the diffuse layer [7, 8].

$$\frac{1}{C_d} = \frac{1}{C_H} + \frac{1}{C_{GC}} \tag{2.4}$$

The dependence of total capacity on potential is shown at the bottom of Figure 2.1(c). E_z is the point where there is zero charge and the capacity is at a minimum.

2.2.4 Modified Theories

Based on these basic EDL theories, the relation between the specific capacitance and surface area would ideally be linear. However, in the practical scenario, there is a considerable deviation from the linear relationship. There have been several reports indicating the contribution of micro and/or sub-micropores in the active surface area and subsequent overall capacitance [10]. Raymundo-Pinero et al. proposed a partial desolvation of solvated ions while entering these pores, which were earlier thought to be inaccessible [11]. A similar trend was also observed by Simon and Gogots and reported an anomalously high capacitance when the size of the pores matches the ionic size. However, correlating this desolvation of ions with the EDL theory is challenging as the confined space within the pores will not be sufficient to accommodate the various layers [12].

In a recent report, Huang et al. developed an approach wherein the pore curvature is considered [13]. Two different models where proposed; an electric double-layer cylinder capacitor (EDCC) and an electric wire in cylinder capacitor (EWCC) model for mesoporous and microporous carbon respectively. When the pore size is sufficiently large so that the pore curvature becomes insignificant, the EDCC model would be brought down to the conventional planar model. The calculation for capacitance according to the EDCC and EWCC models are as follows:

$$C = \frac{\varepsilon_r \varepsilon_0}{b_{\ln}\left[b/(b-d)\right]} A \tag{2.5}$$

$$C = \frac{\varepsilon_r \varepsilon_0}{b_{\ln}\left[b/a_0\right]} A \tag{2.6}$$

It is also to be noted that, by fitting the results, the dielectric constant could be calculated as close to the value at vacuum indicating the fact that the desolvation must be happening before getting inside the micropores [13].

2.3 THEORETICAL TREATMENTS AND MODELLING

Owing to their contrastive charge storage mechanism compared with dielectric capacitors, the conventional models depicting capacitor behaviour are incapable of rightfully representing EDLCs. Hence there are different theoretical models that would represent EDLCs wherein the electrolyte resistance and the EDL capacitance are spread deeper into the pores of the highly porous electrode materials. Hence the theoretical modelling of a supercapacitor would usually consist of capacitor and resistors differing from those of a parallel plate capacitor that gets charged and discharged linearly.

2.3.1 THE CLASSICAL EQUIVALENT CIRCUIT

This is the most simplistic model corresponding with the EDLC for practical application avoiding the complex and undetermined parameters which would actually be there to represent the true physical nature of the system [14]. According to this model, (Figure 2.2(a)), the EDLC is represented with a capacitance (C) with an equivalent series resistance (ESR) and an equivalent parallel resistance (EPR) wherein the EPR and ESR model the ohmic loss and the leakage current during long-term cycling respectively. However, the model is valid only for very low-frequency applications and modified models are required to accurately simulate the system in wide range of frequency [15].

2.3.2 THE THREE-BRANCH MODEL

The three-branch model usually consists of three branches with each of them having a distinct time constant. The charge redistribution as the electrode/electrolyte

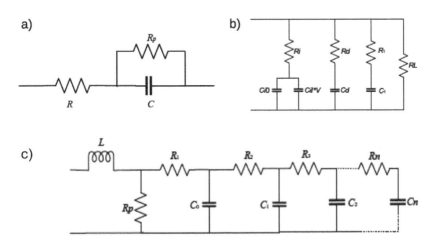

FIGURE 2.2 The Classical Equivalent circuit (a), Three Branch (b) and Transmission Line (c) models representing EDLCs.

interface extends through the pores is also considered (Figure 2.2(b)) [16]. The first branch (denoted as R_i, the immediate branch), has the lowest time constant and it influences the charge/discharge behaviour happening in the order of a few seconds. A nonlinear capacitance is included which would vary as a function of voltage so that the dependence of capacitance with bias voltage is taken into account. Hence, it is modelled as a voltage-dependent differential capacitor consisting of a fixed capacitance and a voltage-dependent capacitor. The second branch is denoted as the delayed branch and here the characteristics in the range of a few minutes are modelled. The third branch is the long term/slow branch influencing the self-discharge behaviour of the circuit when the time is longer than ten minutes. Parallel to the RC branches, a leakage resistor is also included (R_{lea}) so as to model the leakage current. The parameters of the circuit are usually determined by the method illustrated by Zubieta wherein the capacitor is subjected to a fast, controlled charge to examine the successive charge/discharge behaviour. It is found that the three-branch model exhibits an acceptable accuracy within the time range of 30 seconds when the voltage levels are above 40% of the rated voltage [16].

2.3.3   THE POROUS ELECTRODE THEORY

The electrode resistance and the double-layer capacitance are largely affected by electrode porosity. The analysis of porous electrodes by de Levie was one the first extensive analyses of this kind and, according to the porous electrode theory developed by the same group, it is assumed that all the pores in the electrode overlay are non-interconnected and each of them are identical and cylindrical [17, 18]. Moreover, the impedance of a single pore in this condition is adequate to model the whole system. Hence, the impedance characteristic of a single cylindrical pore of length l can be expressed by the electrolyte resistance (R_{el}) which is homogeneously distributed and the double layer capacitance (C_{dl}).

In the assumption that R_{el} and C_{dl} are expressed per unit length of the cylindrical pore, the pore impedance could be calculated from the current and voltage.

$$Z_p \sqrt{\frac{R_{el}}{j \cdot w \cdot c_{dl}}} \coth\left(\sqrt{j \cdot w \cdot C_{dl} R_{el}} \cdot l\right) \tag{2.7}$$

2.3.4   TRANSMISSION LINE MODEL (TLM)

Similar to the transmission line model of traditional capacitors, a transmission line topology also exists for the EDLCs considering the porous nature of the electrodes and the nonlinear relationship of their capacitance with the surface area (Figure 2.2(b)). It is also considered as the simplest approach to apply the impedance of pores. In the model, each of the pores is modelled as a transmission line, modelling a distributed electrolyte resistance as well as a double layer extending deeper into the pores. A perfectly conducting electrode and straight, cylindrical and uniformly sized pores are assumed so as to obtain the capacitive effects [19]. The relaxation time constant depends upon the pore width, the capacitance and resistance of electrolyte in such a way that:

$$\tau = \frac{1}{4}\left(z^2 RC\right) \tag{2.8}$$

Hence, upon application of a DC potential to the pore opening, the penetration depth increases with increase in time and when an AC potential is applied, the penetration depth increases with a decrease in frequency.

The TLM is highly advantageous, firstly because the distributed nature of the supercapacitor is caught physically and it considers the dynamic as well as long-time behaviours. Moreover, simulation is possible over a wide range of frequencies and other simulation software and/or tools can also be implemented [20].

2.4 ENERGY DENSITY AND POWER DENSITY

As in the case of any storage system, energy and power densities are the two major figures of merit for supercapacitor devices. They are very well known for their ultra-high-power profile which is the energy delivered as a function of time which in turn is determined by the maximum voltage and ESR [21]. The right combination of ESR and load gives the highest power density. A compared to typical lithium-ion (Li-ion) batteries and fuel cells, supercapacitors have a much lower energy density which is in the range of 5–10 Wh/kg while that of Li-ion cells and compressed hydrogen are in the ranges of 80–150 Wh/kg and 39,444 Wh/kg respectively [22].

2.4.1 THE RAGONE PLOT

The Ragone plot is the typical representation of energy storage capacity of the system as a function of power output and it compares various storage systems in terms of their energy and power performances [23]. In a representative Ragone plot, energy discharged in watt-hours is shown with respect to the power discharged in watts. However, to represent the actual obtainable power and energy this representation will not be sufficient, and a plot of energy density vs power density would provide more insights into the actual scenario.

The major parameter that determines the Ragone plot is the load resistance, R_L, i.e., the plot represents energy and power that can be dissipated with the variation in load resistance. In order to better understand the relations, a parameter α is introduced in such a way that $\tau = \alpha.\tau o$ and $R_L + R_S = \alpha.R_S$. If we assume R_L to be constant, at time $t = 0$, the initial peak power can be represented as:

$$P_{\alpha,0} = U_0^2 (\alpha-1)/\alpha^2 / R_S \tag{2.9}$$

When $R_S = R_L$ and at time $t = 0$, we can obtain the maximum peak power and is given by:

$$P_{max,0} = E_{max}/2\tau_0 = U_0^2/4\,R_S \tag{2.10}$$

The available part of the stored energy can be obtained as the load power integration as follows:

$$E_\alpha = E_{max}\left(\alpha - 1\right)/\alpha \qquad (2.11)$$

For a particular given R_L, the Ragone plot coordinates will be as follows:

$$E'_\alpha = E_\alpha/M = E_{max}\left(\alpha - 1\right)/\alpha/M \qquad (2.12)$$

$$P'_{\alpha,0} = P_{\alpha,0}/M = 2E_{max}\left(\alpha - 1\right)/\alpha^2/\tau_0/M$$
$$= 4P_{max}\left(\alpha - 1\right)/\alpha^2/M \qquad (2.13)$$

With a load resistance lower than the internal resistance, even with the high current, the voltage drop would be significantly reduced on the load. Subsequently, the power delivered to the load will be small. It is evident from the Ragone plot that, on the right side, there is an interruption on the curve at $P_{max,0}$, the point at which the load resistance equals the internal resistance of the device. The left side of the plot represents the lower power regime wherein the load resistance is considerably greater than the internal resistance [24, 25].

2.5 ELECTRODE MATERIALS

2.5.1 ACTIVATED CARBON

Activated carbon is the most common type of disordered porous carbon. Organic precursors that are rich in carbon, such as coconut shells, wood, leaves, natural fibres, pitch, coal, petroleum, coke, etc., are the natural sources of activated carbon. Activated carbon is produced from these organic precursors through two steps: the first step is the process of carbonization which involves the heat treatment of organic precursors under an inert atmosphere and in the second step carbon precursor grains are activated by partial controlled oxidation through a physical or chemical method [26]. The process of activation helps to get a porous network, resulting in the formation of carbon particles with a high surface area. Activation can be done either physically or chemically. In physical activation, carbon precursor grains are subjected to carbonization at high temperature (700–1200°C) under an oxidizing atmosphere (in the presence of H_2O, CO_2 or air). Production of activated carbon through a chemical method involves treating the carbon precursor grains with chemical activating reagents such as KOH, H_2SO_4, $ZnCl_2$, etc. A wide range of pore structures is present in activated carbon, ranging from macropores (>50 nm), mesopores (2–50 nm) and micropores (<2 nm). The electrochemical properties of an activated carbon electrode are influenced by various parameters such as its surface area, the pore structure, the pore size distribution and the surface functionality. Sometimes pore volume becomes large as a result of excessive activation, which may lead to the loss of desired electrochemical properties. It is possible to produce activated carbon with more advanced properties such as well-developed porosity and more uniform pores and microstructure by using synthetic polymers (such as poly aniline, poly vinyl chloride, polyfurfuryl alcohol, etc.) as the precursors. It was observed that an activated carbon-based supercapacitor electrode delivers higher capacitance in aqueous electrolytes (100–300 F/g) than organic electrolytes (<150 F/g) [27–29].

2.5.2 Carbon Nanotubes (CNTs)

CNTs have been widely studied as an electrode material owing to their high conductivity, good thermal and mechanical stability and unique pore structure. They have a tubular structure and are most commonly grown through chemical vapor deposition methods assisted by catalysts. Hydrocarbon-based gaseous precursors such as methane, acetylene, propylene and others are commonly used. CNTs are of two types; single walled nanotubes (SWNTs) and multi walled nanotubes (MWNTs). Both are explored as an electrode material for supercapacitors [30, 31]. SWNTs are highly flexible, but not easy to handle due to their tendency to form bundles. Compared to SWNTs, MWNTs have a number of defects, which would result in a deviation from their structure as well as their intrinsic properties, which in turn may lead to higher resistance. MWNTs are longer in size and have a greater volume than SWNTs. CNTs are highly conductive, and compared to other conductors there is no heating up while conducting electricity (Joule heating). This is because conduction in CNTs is through ballistic transport, i.e., charge carriers have a very high mean free path and are not subjected to scattering, thus eliminating Joule heating. A major limitation with CNTs is their low specific surface area compared to activated carbon. The reported specific capacitance of pure CNTs is less than 200 F/g. It is possible to improve charge storage properties of CNTs either by functionalizing or by compositing with other components [32].

2.5.3 Carbon Aerogel

Carbon aerogel is a highly porous and the lightest material obtained from a gel in which the liquid component of the gel is replaced with a gas. The sol-gel method is used to produce carbon aerogels by the polycondensation reaction of formaldehyde and resorcinol. A uniform mesoporous structure is obtained when pyrolyzing in a nitrogen atmosphere. The specific surface area of carbon aerogels ranges from 100 to 1000 m^2/g and has demonstrated specific capacitance value of <50 F/g) (in KOH and H_2SO_4 electrolytes) [33]. Activation and introduction of functional groups on the surface of carbon aerogel helps to enhance the specific capacitance value up to 200 F/g [34].

2.5.4 Carbon Nanofibre (CNF)

CNFs are graphitized one-dimensional carbon nanostructures and they have a tower-like morphology that arises as a result of the stacking of very small structural basic units of graphene [35]. CNFs can be synthesized through various methods like electrospinning, chemical vapor deposition, template-assisted solvothermal method, pyrolysis of biomaterial, etc., among which electrospinning is the most widely used. The processing method largely influences the properties of CNF. The electrochemical properties of CNF can be enhanced by various methods such as adding a pseudocapacitive material, incorporating a highly conductive material, manipulating the fibre morphology as well as the degree of graphitization or activation with a hetero atom such as phosphorous, boron or nitrogen. For instance, a specific capacitance of 175 Fg^{-1} was obtained when nitrogen was doped into the cross-linked carbon network and an energy density and power density of 5.9 $WhKg^{-1}$ and 242.2 $WhKg^{-1}$ respectively, was delivered by assembled supercapacitor [36]. The electrical conductivity of CNF can be enhanced by incorporating CNTs or metal nanoparticles.

2.5.5 GRAPHENE

Graphene is a 2D allotrope of carbon. It is a one-atom-thick layer and other dimensionalities like 0D Bucky balls, 1D nanotube or 3D graphite are obtained by wrapping, rolling or stacking respectively. The properties of graphene such as high specific surface area ($\approx$2630 m^2/g) and high intrinsic specific capacitance ($\approx$550 Fg^{-1}) make it a suitable electrode material for double-layer capacitors [37]. Chemical vapour deposition, mechanical exfoliation, arc discharge method, chemical derivation method, etc., are some of the popular techniques used for the synthesis of graphene. The specific capacitance value of graphene is strongly influenced by factors such as the method adopted for synthesis, type of electrolyte used and method of electrode design. High flexibility, compatible mechanical properties and low ion transport diffusion resistance renders graphene as a preferred electrode for flexible supercapacitors. But the major disadvantage associated with graphene is its tendency of agglomeration. The restacking of single graphene sheets limits the electrolyte ion diffusion which further deteriorates the electrochemical performance [38]. This problem can be overcome by introducing a spacer, which helps to increase the space between the graphene layers. These spacers can be made from transition metal oxides, conducting polymers or carbon materials. By incorporating carbon black nanoparticles as a spacer, Wang et al. synthesized graphene paper electrode which delivered a higher capacitance (138 Fg^{-1} at 10 mV/s) than pure graphene (83.2 F/g) [39].

2.5.6 FULLERENE

Fullerenes are carbon balls with pentagonal and hexagonal rings of carbon. They have a small band gap and a high conductivity which allows fast charge transfer. Fullerenes have been used to improve the performance of supercapacitor. Incorporation of C_{60} in graphene sheets enhances the charge transfer property of the electrodes. Compositing fullerenes with transition metal oxides, conducting polymers and layered chalcogenides helps to favourably modify the electron transfer and ion diffusion property of the electrodes [40–42].

Besides the aforementioned materials, hierarchical porous carbon, carbide-derived carbon, onion-like carbon, mesoporous carbon, etc., are some examples of popular carbon-based electrode materials [43–46]. The advantages of carbon-based materials include higher power performance, environmental friendliness, low processing cost, high temperature tolerance, etc. But the major disadvantage is their limited specific capacitance value which is insufficient for practical applications. However, the specific capacitance value delivered by pseudocapacitive electrode materials (such as metal oxides and conducting polymers) is almost ten to 100 times higher than that of carbon-based EDLC materials. Therefore, researchers have been trying to explore more pseudocapacitive electrode materials for designing electrochemical capacitors.

2.6   ELECTROLYTE MATERIALS

Electrolytes have a pivotal role in determining the efficiency of EDLCs, as the charge storage in EDLCs occurs via the adsorption of electrolyte ions on the electrode surface and are hence considered to be an active part of the system. Parameters like viscosity, electrical conductivity and electrochemical stability are especially crucial, influencing the ultimate performance of the system. In order to obtain a high-power profile for the device, low viscosity and high conductivity are required which would impart a faster ion movement and high power. On the other hand, an electrolyte with high electrochemical stability provides high operating voltages which improve the energy density as well as power density of the system [47]. The dielectric constant, melting/boiling point and salt chemistry of the electrolytes also play key roles in determining the ultimate performance of the device. Hence formulation of an optimized electrolyte satisfying all these criteria is really challenging. Another major factor to be considered is the toxicity of the electrolyte as well as the decomposition products, if any, to meet the chemical safety requirements.

Aqueous electrolytes of acidic, alkaline or neutral types are widely used in the laboratory scale EDLC prototypes, owing to their low internal resistance and viscosity; however, they do suffer from low electrochemical stability which limits the potential window which in turn would limit the energy as well as power density [48]. Hence the recent trend in EDLC development, especially in the case of commercial EDLCs, is to switch from aqueous electrolytes to organic electrolytes which would allow an operating voltage up to 2.5–3 V and have a moderate viscosity and conductivity [49]. Aprotic solvents like acetonitrile or carbonate-based solvents are typically being used in this regard. However, there are several disadvantages including the complex purification processes required, high flammability, toxicity, rising cost and safety concerns. Moreover, the lower conductivity and low dielectric constant would affect the power profile and capacitance values respectively [47, 49].

Another broad class of potential electrolyte material include ionic liquids, especially due to the wide operating voltage and also the various combinations of anions and cations. Commercial devices with a rating as high as 1000 F could be achieved with optimal cation and anion choice. However, the utilization of ionic liquid-based electrolytes is limited to high temperature applications owing to their low conductivity (a few millisiemens per cm) [49]. For instance, carbon-based electrodes in combination with a 1-Ethyl-3-methylimidazolium/bis(trifluoromethylsulfonyl) imide (EMI/TFSI) ionic liquid (IL) electrolyte could exhibit a capacitance of 160 Fg^{-1} (and 90 Fcm^{-3}) at 60°C [50]. Nevertheless, at the lower temperature region, (up to 60°C), where conventional Li-ion batteries and supercapacitor work, ILs fail to meet the standards owing to this low conductivity. However extensive research is ongoing to optimize a suitable combination of anion and cation and also the right electrolyte additives or organic solvents so that ILs do become a potential candidate for future storage devices [51, 52].

2.7   SUMMARY AND OUTLOOK

Supercapacitors have been considered to be the most fascinating storage technology for the past few decades. EDLCs can be considered to be the first generation of supercapacitors and in spite of the booming of electrode innovations in terms of materials as well as design, porous carbon-based EDLCs remain the leading candidates in practical applications. The chapter covers the basic principles and theories of EDLCs and also the current trends in choosing the electrode and electrolyte materials to improve the efficiency of the system. Trends in the energy and power performance of the system and the factors that determine these figures of merit are also discussed.

REFERENCES

1. Sharma, P., & Bhatti, T. S. (2010). A review on electrochemical double-layer capacitors. *Energy Conversion and Management, 51*(12), 2901–2912.
2. Yu, A., Chabot, V., & Zhang, J. (2013). *Electrochemical Supercapacitors for Energy Storage and Delivery: Fundamentals and Applications* (p. 383). Taylor & Francis.
3. Li, T., Ma, R., Lin, J., Hu, Y., Zhang, P., Sun, S., & Fang, L. (2020). The synthesis and performance analysis of various biomass-based carbon materials for electric double-layer capacitors: A review. *International Journal of Energy Research, 44*(4), 2426–2454.
4. Nishihara, H., Itoi, H., Kogure, T., Hou, P. X., Touhara, H., Okino, F., & Kyotani, T. (2009). Investigation of the ion storage/transfer behavior in an electrical double-layer capacitor by using ordered microporous carbons as model materials. *Chemistry: A European Journal, 15*(21), 5355–5363.
5. Bullard, G. L., Sierra-Alcazar, H. B., Lee, H. L., & Morris, J. L. (1989). Operating principles of the ultracapacitor. *IEEE Transactions on Magnetics, 25*(1), 102–106.
6. Endo, M., Takeda, T., Kim, Y. J., Koshiba, K., & Ishii, K. (2001). High power electric double layer capacitor (EDLC's); from operating principle to pore size control in advanced activated carbons. *Carbon Letters, 1*(3_4), 117–128.
7. Del Gratta, C., & Romani, G. L. (1999). MEG: Principles, methods, and applications. *Biomedizinische Technik/Biomedical Engineering, 44*(s2), 11–23.
8. Aulice Scibiohb, B. V. M., *Materials for Supercapacitor Applications, I*, Elsevier, 2020.
9. Kar, K. K. (2020). *Handbook of Nanocomposite Supercapacitor Materials II*. Springer International Publishing.

10. Shao, H., Wu, Y. C., Lin, Z., Taberna, P. L., & Simon, P. (2020). Nanoporous carbon for electrochemical capacitive energy storage. *Chemical Society Reviews, 49*(10), 3005–3039.

11. Raymundo-Pinero, E., Kierzek, K., Machnikowski, J., & Béguin, F. (2006). Relationship between the nanoporous texture of activated carbons and their capacitance properties in different electrolytes. *Carbon, 44*(12), 2498–2507.

12. Gogotsi, Y., & Simon, P. (2011). True performance metrics in electrochemical energy storage. *science, 334*(6058), 917–918.

13. Huang, J., Sumpter, B. G., & Meunier, V. (2008). A universal model for nanoporous carbon supercapacitors applicable to diverse pore regimes, carbon materials, and electrolytes. *Chemistry–A European Journal, 14*(22), 6614–6626.

14. Nelms, R. M., Cahela, D. R., & Tatarchuk, B. J. (2003). Modeling double-layer capacitor behavior using ladder circuits. *IEEE Transactions on Aerospace and Electronic Systems, 39*(2), 430–438.

15. Spyker, R. L., & Nelms, R. M. (2000). Classical equivalent circuit parameters for a double-layer capacitor. *IEEE Transactions on Aerospace and Electronic Systems, 36*(3), 829–836.

16. Zubieta, L., & Bonert, R. (2000). Characterization of double-layer capacitors for power electronics applications. *IEEE Transactions on Industry Applications, 36*(1), 199–205.

17. De Levie, R. (1963). On porous electrodes in electrolyte solutions: I. Capacitance effects. *Electrochimica Acta, 8*(10), 751–780.

18. De Levie, R. (1964). On porous electrodes in electrolyte solutions—IV. *Electrochimica Acta, 9*(9), 1231–1245.

19. Cultura, A. B., & Salameh, Z. M. (2015). Modeling, evaluation and simulation of a supercapacitor module for energy storage application. *Cell, 1*(1), 1.

20. Dougal, R. A., Gao, L., & Liu, S. (2004). Ultracapacitor model with automatic order selection and capacity scaling for dynamic system simulation. *Journal of Power Sources, 126*(1–2), 250–25.

21. Yassine, M., & Fabris, D. (2017). Performance of commercially available supercapacitors. *Energies, 10*(9), 1340.

22. Gidwani, M., Bhagwani, A., & Rohra, N. (2014). Supercapacitors: The near future of batteries. *International Journal of Engineering Inventions, 4*(5), 22–2.

23. Christen, T., & Carlen, M. W. (2000). Theory of Ragone plots. *Journal of Power Sources, 91*(2), 210–216.

24. Schneuwly, A., & Gallay, R. (2000, June). Properties and applications of supercapacitors from the state-of-the-art to future trends. In *Proceeding PCIM* (Vol. 2000).

25. Namisnyk, A., & Zhu, J. (2003). A survey of electrochemical super-capacitor technology. In *Australian Universities Power Engineering Conference*. University of Canterbury, New Zealand.

26. Poulomy Roy, S. S., *Nanomaterials for Electrochemical Energy Storage Devices*, Wiley, 2019.

27. Najib, S., & Erdem, E. (2019). Current progress achieved in novel materials for supercapacitor electrodes: Mini review. *Nanoscale Advances, 1*(8), 2817–2827.

28. Gu, W., & Yushin, G. (2014). Review of nanostructured carbon materials for electrochemical capacitor applications: Advantages and limitations of activated carbon, carbide-derived carbon, zeolite-templated carbon, carbon aerogels, carbon nanotubes, onion-like carbon, and graphene. *Wiley Interdisciplinary Reviews: Energy and Environment, 3*(5), 424–473.

29. Iro, Z. S., Subramani, C., & Dash, S. S. (2016). A brief review on electrode materials for supercapacitor. *International Journal of Electrochemical Sciences, 11*(12), 10628–10643.

30. Kanninen, P., Luong, N. D., Anoshkin, I. V., Tsapenko, A., Seppälä, J., Nasibulin, A. G., & Kallio, T. (2016). Transparent and flexible high-performance supercapacitors based on single-walled carbon nanotube films. *Nanotechnology, 27*(23), 235403.

31. Pandey, G. P., Hashmi, S. A., & Kumar, Y. (2009). Multiwalled carbon nanotube electrodes for electrical double layer capacitors with ionic liquid-based gel polymer electrolytes. *Journal of The Electrochemical Society, 157*(1), A105.

32. Boyea, J. M., Camacho, R. E., Sturano, S. P., & Ready, W. J. (2007). Carbon nanotube-based supercapacitors: Technologies and markets. *Nanotechnology Law & Business, 4*, 19.

33. Li, J., Wang, X., Huang, Q., Gamboa, S., & Sebastian, P. J. (2006). Studies on preparation and performances of carbon aerogel electrodes for the application of supercapacitor. *Journal of Power Sources, 158*(1), 784–788.

34. Kalpana, D., Renganathan, N. G., & Pitchumani, S. (2006). A new class of alkaline polymer gel electrolyte for carbon aerogel supercapacitors. *Journal of Power Sources, 157*(1), 621–623.

35. Su, D. S., & Schlögl, R. (2010). Nanostructured carbon and carbon nanocomposites for electrochemical energy storage applications. *ChemSusChem: Chemistry & Sustainability Energy & Materials, 3*(2), 136–168.

36. Cheng, Y., Huang, L., Xiao, X., Yao, B., Yuan, L., Li, T., ... & Zhou, J. (2015). Flexible and cross-linked N-doped carbon nanofiber network for high performance freestanding supercapacitor electrode. *Nano Energy, 15*, 66–74.

37. Yang, W., Ni, M., Ren, X., Tian, Y., Li, N., Su, Y., & Zhang, X. (2015). Graphene in supercapacitor applications. *Current Opinion in Colloid & Interface Science, 20*(5–6), 416–428.

38. Wang, Y., Wu, Y., Huang, Y., Zhang, F., Yang, X., Ma, Y., & Chen, Y. (2011). Preventing graphene sheets from restacking for high-capacitance performance. *The Journal of Physical Chemistry C, 115*(46), 23192–23197.

39. Wang, M. X., Liu, Q., Sun, H. F., Stach, E. A., Zhang, H., Stanciu, L., & Xie, J. (2012). Preparation of high-surface-area carbon nanoparticle/graphene composites. *Carbon, 50*(10), 3845–3853.

40. Ali, B. A., Biby, A. H., & Allam, N. K. (2020). Fullerene C76: An unexplored superior electrode material with wide operating potential window for high-performance supercapacitors. *ChemElectroChem, 7*(7), 1672–1678.

41. Benzigar, M. R., Joseph, S., Saianand, G., Gopalan, A. I., Sarkar, S., Srinivasan, S., ... & Vinu, A. (2019). Highly ordered iron oxide-mesoporous fullerene nanocomposites for oxygen reduction reaction and supercapacitor applications. *Microporous and Mesoporous Materials, 285*, 21–31.

42. Xiong, S., Yang, F., Jiang, H., Ma, J., & Lu, X. (2012). Covalently bonded polyaniline/fullerene hybrids with coral-like morphology for high-performance supercapacitor. *Electrochimica Acta, 85*, 235–242.

43. Qie, L., Chen, W., Xu, H., Xiong, X., Jiang, Y., Zou, F., Huang, Y. (2013). Synthesis of functionalized 3D hierarchical porous carbon for high-performance supercapacitors. *Energy & Environmental Science, 6*(8), 2497–2504.

44. Gao, P. C., Tsai, W. Y., Daffos, B., Taberna, P. L., Pérez, C. R., Gogotsi, Y., Favier, F. (2015). Graphene-like carbide derived carbon for high-power supercapacitors. *Nano Energy, 12*, 197–206.

45. Pech, D., Brunet, M., Durou, H., Huang, P., Mochalin, V., Gogotsi, Y., Simon, P. (2010). Ultrahigh-power micrometre-sized supercapacitors based on onion-like carbon. *Nature Nanotechnology, 5*(9), 651–654.

46. Prabaharan, S. R. S., Vimala, R., & Zainal, Z. (2006). Nanostructured mesoporous carbon as electrodes for supercapacitors. *Journal of Power Sources, 161*(1), 730–736.

47. Pal, B., Yang, S., Ramesh, S., Thangadurai, V., & Jose, R. (2019). Electrolyte selection for supercapacitive devices: A critical review. *Nanoscale Advances, 1*(10), 3807–3835.

48. Zhang, X., Wang, X., Jiang, L., Wu, H., Wu, C., & Su, J. (2012). Effect of aqueous electrolytes on the electrochemical behaviors of supercapacitors based on hierarchically porous carbons. *Journal of Power Sources, 216*, 290–296.

49. Kurzweil, P., & Chwistek, M. (2008). Electrochemical stability of organic electrolytes in supercapacitors: Spectroscopy and gas analysis of decomposition products. *Journal of Power Sources, 176*(2), 555–567.

50. Yuyama, K., Masuda, G., Yoshida, H., & Sato, T. (2006). Ionic liquids containing the tetrafluoroborate anion have the best performance and stability for electric double layer capacitor applications. *Journal of Power Sources, 162*(2), 1401–1408.

51. Thompson, M. W., Matsumoto, R., Sacci, R. L., Sanders, N. C., & Cummings, P. T. (2019). Scalable screening of soft matter: A case study of mixtures of ionic liquids and organic solvents. *The Journal of Physical Chemistry B, 123*(6), 1340–1347.

52. Kühnel, R. S., Obeidi, S., Lübke, M., Lex-Balducci, A., & Balducci, A. (2013). Evaluation of the wetting time of porous electrodes in electrolytic solutions containing ionic liquid. *Journal of Applied Electrochemistry, 43*(7), 697–704.

3 Fundamentals of Pseudocapacitors

Shyamli Ashok, Deepu Joseph,
Jasmine Thomas and Nygil Thomas

CONTENTS

DOI: 10.1201/9781003258384-3

3.1 INTRODUCTION

As we prepare to witness the miracles of the Fourth Industrial Revolution where hyper-connectivity decides the pace of transformation, the design and development of efficient, powerful and better energy storage devices such as supercapacitors becomes an urgent necessity. The intermittent nature of renewable resources such as wind, solar energy, hydrothermal and geothermal energy necessitates fast and efficient energy storage techniques so as to meet global energy demands. Moreover, concern about the energy crisis coupled with environmental issues has impelled researchers to develop efficient and sustainable energy storage devices capable of delivering high power and energy densities [1–3]. Even though the first-generation electric double-layer supercapacitors are expected to augment the high energy density of conventional batteries with their high power density characteristics, their low energy density remains a challenge in practical applications [4–6]. This is especially so when we try to further expand the technology so as to replace the conventional storage devices [7, 8]. The development of pseudocapacitors addresses this concern to a large extent as they exploit the redox processes at the electrode surface, improving the storage capability and capacitance value, which in turn would also improve the energy density.

3.2 PSEUDOCAPACITORS

A supercapacitor device is mainly composed of four components – an active material, suitable electrolyte, current collector and a separator. Among these, active material plays a vital role in determining the performance of a supercapacitor device. Based on the type of active material and the mechanism of charge storage in them, supercapacitors are categorized into three: (1) electric double-layer capacitors (EDLCs), (2) pseudocapacitors and (3) hybrid capacitors. Charge storage in EDLCs is facilitated by the formation of an electrical double layer at the electrode/electrolyte interface. It involves a non-faradaic process where no charge transfer happens between the electrode and the electrolyte. When an external voltage is applied, electrolyte ions are accumulated on the surface of the oppositely charged electrode as a result of electrostatic attraction. EDLCs permit better power performance and withstand several charge/discharge cycles. However, owing to the lower energy density of EDLCs, several other electrode processes are included in the system to improve the charge storage as well as energy performance, which together are called pseudocapacitive processes. Pseudocapacitance is purely faradaic in origin, where the process of charge storage is a result of highly reversible and fast redox reactions at or near the surface. The difference of pseudocapacitive material from a battery type is that the kinetics of the former is limited by surface-related processes [9]. The concept of pseudocapacitance was first used by Conway and Gileadi in 1962 where the electrochemical adsorption of chemical species like O, OH or H ions on electrode surfaces were found to be associated with a reversible capacitance. The capacitive electrochemical performance of hydrous RuO_2 was then reported in 1971 by Trasatti and Buzzanca, which spurred extensive research in the field of various pseudocapacitor materials. The redox reactions in pseudocapacitors are compared to batteries and are characterized by faradaic reactions wherein charges move across the double layer, similar to what happens during the charge/discharge mechanism in batteries. For pseudocapacitors, the derivative $\dfrac{\Delta q}{\Delta V}$ or $\dfrac{dq}{dV}$, where Δq is the charge acceptance and ΔV is the change of potential, gives the capacitance.

Pseudocapacitive behaviour of a material can be either extrinsic or intrinsic. An extrinsic pseudocapacitive material exhibits pseudocapacitive behaviour only when its size is reduced to nanoscale and not in bulk, whereas an intrinsic pseudocapacitive material exhibits its pseudocapacitive property irrespective of the particle size and morphology. Metal oxides and conducting polymers are the commonly used pseudocapacitive materials [10]. Hybrid capacitors combine both faradaic and non-faradaic types of charge storage. Combining the advantages from both, results in high energy and power density, better rate capability and improved cycle life [11, 12].

Depending upon the processes involved, electrochemical energy storage originates by three different means: (i) through redox reactions at the surface, (ii) through the intercalation/deintercalation of ions and (iii) conversion, decomposition or alloying reactions [12]. All these three processes are considered to be of faradaic origin as they proceed via a charge transfer reaction across electrode/electrolyte interface and Faraday's laws are obeyed. A deeper insight into the electrode processes is beyond the scope of this chapter and is discussed in Chapter 4.

3.3   ENERGETICS AND KINETICS OF PSEUDOCAPACITANCE

Out of the three different faradaic processes discussed in the previous section, the surface redox reactions and ion intercalation/deintercalation are termed as pseudocapacitive since the energetics and thermodynamics of these processes can be clearly depicted by the same mathematical models as those of the surface adsorption/desorption [13]. Irrespective of the mechanism involved, the extent of reaction ξ has a linear (or almost linear) dependence with the potential V, so that the simplistic thermodynamic definition of pseudocapacitance would be as $\Delta \xi / \Delta V$. It was also proposed that the surface adsorption and coverage of these species have a linear relation with the heat of adsorption [13, 14]. The concept was further modified by Conway to fit with monolayer adsorption [15].

When we consider the electrosorption of a cation C^+ onto the surface of a substrate S,

$$S + C^+ + e^- \leftrightarrow Sc_{ads} \tag{3.1}$$

considering the adsorbed species to follow a Langmuir electrosorption isotherm,

$$\frac{\theta}{(1-\theta)} = KC_A e^{VFRT} \tag{3.2}$$

where C_A is the concentration f cation in the solution, θ is the surface coverage of Sc_{ads}, $(1 - \theta)$ is the surface coverage of S, K is the ratio of forward and reverse reaction rate constant, V is the electrode potential, T is the temperature and F and R are Faraday constant and the ideal gas constant respectively [15].

Rearranging the equation to adopt the form of a Nernst equation (K is replaced by the extent of adsorbate surface coverage) gives:

$$E = E^o + \frac{RT}{F} \ln\left[\frac{1.\theta}{KC_A(1-\theta)}\right] \tag{3.3}$$

Here, the equilibrium potential of the redox couple is termed as E, whereas E^o represents the standard potential of the redox process. Assuming that a charge q is required to complete a monolayer coverage, C_{ads}, the pseudocapacitance is represented as follows:

$$C_\Phi = q\frac{d\theta}{dV} = \frac{qF}{RT}\frac{KC_Ae^{VFRT}}{\left(1+KC_Ae^{VFRT}\right)^2} = \frac{qF}{RT}\theta(1-\theta) \tag{3.4}$$

It is evident from Equation (3.4) that when $\theta = \frac{1}{2}$, the pseudocapacitance reaches its maximum and hence, for perfect Langmuir-type electrosorption, the pseudocapacitance is a function of voltage applied and surface coverage [15]. According to this formalism, the potential dependence of capacitance holds good even with multiple redox reactions; however, there are also materials that may lack distinct faradaic peaks. In order to accommodate the various material types, Costentin et al. suggested a combined EDLC and pseudocapacitor mechanism with the assumption that the electrochemically active surface area of the electrodes must be greater than the physically measured surface area so that there is a bigger contribution from EDLC mechanism [16]. Conway further extended his concept to intercalation as well as redox active electrodes by replacing the surface coverage, θ, with lattice occupancy and concentration of oxidant respectively so that the dependence of applied potential is the only significant factor. He established the fact that the critical feature of any pseudocapacitive material is the reversibility of the process over a wide time domain with no limitations based on mass transfer and the absence of significant phase transformation reactions like conversion or alloying that would happen in the case of batteries [17].

3.3.1 THE DEFINITION OF PSEUDOCAPACITANCE

In spite of these kinetic and energetic formulations there has been an obvious evolution in the concepts, understanding and in fact the actual definition of pseudocapacitance as these atomistic descriptions, quite often, could not have extended to a wide range of materials and processes. For instance, Brousse et al. formulated a more confined definition which is valid only for materials with a constant capacitance over a wide potential range, so as to exclude materials like NiO in alkaline electrolytes being wrongly reported as a pseudocapacitive material with huge capacitance values [18]. In another report, Simon et al. suggested a new nomenclature, extrinsic pseudo-capacitance for nanostructured materials. It was proposed that when the intercalation reaction does not involve a significant mass transfer, the diffusion length becomes smaller than the square root of the diffusion coefficient multiplied by diffusion time [19]. Moreover, as the number of redox sites at the electrode/electrolyte interface constitutes a major fraction of the total redox sites in nanostructured materials, the more dispersed sites across the interface cause a more capacitor-like electrochemical performance. There are also several approaches where the surface adsorption and intercalation pseudocapacitance becomes mechanically similar [20]. In short, the actual interpretation of the concept "pseudocapacitance" from a mechanical per-spective, remains a major lacuna in the field of electrochemical energy storage.

3.4 ELECTRODE MATERIALS

The nature of the electrode material is the most critical factor that decides the perfor-mance of a supercapacitor; great emphasis is laid in the synthesis and optimization of suitable candidates for the purpose. A wide spectrum of materials synthesized through a variety of routes has been in use and the search for novel materials with exotic performance is underway. There is a broad spectrum of materials used as pseudocapacitor electrodes and the major classes of such materials are as follows.

3.4.1 Metal Oxides

Metal oxides are potential candidates for fabricating supercapacitor electrodes due to their high conductivity and large specific capacitance. They have a better charge storage efficacy than traditional carbon-based material due to the fast reversible fara-daic reaction occurring at the surface and near the surface region. Specific surface area, wettability, crystallinity and particle size are the prime factors that influence the electrochemical performance of metal oxide-based electrode materials. RuO_2, MnO_2, NiO and vanadium oxides are some of the most commonly used metal oxides.

3.4.1.1 Ruthenium Oxide (RuO_2)

RuO_2 is the first reported pseudocapacitive material in the literature. The estimated theoretical specific capacitance value for RuO_2 is 1400–2000 Fg^{-1} [21, 22]. It shows high electrochemical reversibility, high conductivity and extensive cyclability. RuO_2 exists either in amorphous or in crystalline form. Amorphous RuO_2 yields a higher specific capacitance than its rutile structure. The charge/discharge process in RuO_2 can be represented as follows:

$$Ru^{IV}O_2 + xH^+ + xe^- \leftrightarrow Ru^{IV}_{1-x}Ru^{III}_x O_2 H_x$$

During this process, a rapid transfer of electrons occurs, which is accompanied by elec-trosorption of a proton at the surface of RuO_2. In nanocrystalline hydrous RuO_2 (RuO_2.

XH_2O), structural water content plays a major role in determining the specific capacitance and conductivity of the electrode material. RuO_2 facilitates electron transfer while the role of water molecules is to provide the path for proton movement that contributes to the charge storage mechanism [21]. There are different methods used for the synthesis of nanocrystalline hydrous RuO_2 which includes sol-gel techniques, atomic layer deposition, anodic deposition techniques and the template-assisted hydrothermal method. A major limitation with RuO_2 electrodes is its high cost and rare resource.

3.4.1.2 Manganese Oxide (MnO₂)

As a result of many attempts to find inexpensive and eco-green materials to substitute for $RuO_2.H_2O$ as supercapacitor electrodes, researchers have found MnO_2 as a potential candidate for supercapacitor electrodes since it is cost effective, environmentally benign and exhibits high pseudocapacitance. During the reversible reactions, the transition between oxidation states (III and IV) takes place. There are different crystallographic forms of MnO_2 such as α, β, γ and δ. The supercapacitive performance of MnO_2 electrodes depends on its morphology, crystallinity, conductivity, etc. Various synthesis routes, as well as various morphologies, were reported for MnO_2. In most of the reported works, $KMnO_4$, $Mn(CH_3COO)_2$ or $MnSO_4$ was used as the Mn precursors irrespective of the synthesis method or morphology [23]. MnO_2 electrodes can deliver a maximum specific capacitance of 1300 Fg^{-1}. The charge storage mechanism in MnO_2 can be explained in two ways. The first case is the intercalation of alkali metal cations like Na, K, etc. (C^+) or protons (H^+) upon reduction and deintercalation of protons and cations upon oxidation [24]. This is shown below:

$$MnO_2 + H^+ + e^- \Leftrightarrow MnOOH \text{ or } MnO_2 + C^+ + e \Leftrightarrow MnOOC \qquad (3.5)$$

The second mechanism is based on the surface adsorption of cations from the electrolyte solution (C^+) on MnO_2. This can be represented as $(MnO_2)_{surface} + C^+ + e^- \Longleftrightarrow (MnO_2^-C^+)_{surface}$.

But MnO_2 electrodes suffer from low conductivity, which can be addressed by compositing MnO_2 with conducting carbons or other metal oxides. Hydrothermal, sol-gel, coprecipitation and simple reduction are the most common methods used for the synthesis of MnO_2.

3.4.1.3 Nickel Oxide (NiO)

NiO is considered as a promising candidate for pseudocapacitors among various other metal oxides, because of its high theoretical capacitance, easy availability, cost effectiveness and good thermal and chemical stability. The electrochemical reactions of NiO in alkaline electrolyte can be expressed as follows [25]:

$$NiO + OH^- \rightleftarrows NiOOH + e^-$$

Various morphologies exhibited by NiO are nanosheets, nanocolumns, nanofibres, nanoflowers, etc. By modifying the method and conditions of synthesis, it is possible to obtain NiO with different morphologies, specific surface area and pore size. The processing method and morphology affect the electrochemical performance of NiO electrodes. However, the experimental specific capacitance value is lower than the theoretical value due to the low electrical conductivity. The conductivity of NiO electrodes can be increased by compositing with carbon-based materials, conducting polymers or other metal oxides. By using the hydrothermal method followed by a thermal treatment, Chongyong et al. synthesized 3D flower-like nickel oxide on graphene sheets. Compared to bare NiO, the composite exhibited a high specific capacitance value (778.7 F/g) [26].

3.4.1.4   Vanadium Oxides

Over the past years, vanadium oxides have been extensively used due to their unique optical, electrical and electrochemical properties. Vanadium monoxide (VO), vanadium dioxide (VO_2), vanadium trioxide (V_2O_3), vanadium pentoxide (V_2O_5) and various other forms like V_3O_7, V_6O_{13} and V_4O_9 are the different compounds that vanadium creates with oxygen due to the variable oxidation states of vanadium. Each of these materials exhibits its own specific properties. Among these, vanadium pentoxide (V_2O_5) and vanadium dioxide (VO_2) are being looked upon as promising cathode materials for electrochemical energy storage due to their high capacity/capacitance, wide potential window and excellent cycling stability [27]. VO_2 is a semiconductor with a low band gap of 0.7 eV. Compared to transition metal oxides, it has relatively high conductivity. At about 68°C, it reversibly changes from monoclinic structure to triclinic structure and undergoes a transition from semiconductor to metal [28]. 2D VO_2 nanosheets were synthesized by Rakhi et al. by hydrothermal reduction of V_2O_5 nanopowders. They tested the electrochemical performance of VO_2 nanosheets by fabricating a symmetric supercapacitor using a liquid organic electrolyte as well as flexible solid-state supercapacitors using alumina-silica-based gel electrolyte. In both devices, only the VO_2 nanosheets were used as the electrode. The former delivered an energy density and power density of 46 Whkg^{-1} and 1.4 kWkg^{-1} respectively and the later exhibited a maximum specific capacitance of 145 Fg^{-1}. It was also possible to light up red LEDs by connecting three solid-state capacitors in series for more than one minute [28]. V_2O_5 has a layered structure with a layer separation of 10–14 Å. This van der Waals gap provides easy path for the diffusion of electrolyte ions.

Besides the above-mentioned transition metal oxides, cobalt oxides (Co_3O_4, CoO_x), zinc oxides (ZnO), iron oxides (Fe_2O_3, Fe_3O_4) and titanium oxide (TiO_2) are some of the commonly used metal oxides for supercapacitor applications.

3.4.2   Binary and Ternary oxides

Binary metal oxides are composed of one transition metal ion and an electrochemically active ion. MCo_2O_4 (M = Mn, Ni, Cu, Zn, etc.), $MMoO_4$ (M = Ni, Co), MFe_2O_4 (M = Ni, Co, Sn, Mn), $MnSO_3$ (M = Ni, Co), $NiMn_2O_4$ are some examples of binary metal oxides which are extensively used as the supercapacitor electrode material. It has advantages over pure metal ions and the synergetic effect from more than one metal ion, which helps to improve electrochemical performance. They show superior conductivity, wide potential window, improved stability and more redox sites. Investigation of ternary metal oxides has also been studied to investigate effect of incorporation of more metals. In this way many different ternary oxides such as Zn-Ni-Co oxide nanowire arrays on nickel foam, flower-like Ni-Zn-Co oxide, nickel manganese ferrite film electrode, etc., have been used as electrode material and could deliver a high specific capacitance value. Huang et al. synthesized a $Ni_xCo_yMo_zO$ electrode which was found to exhibit better performance than Ni_xMo_yO, Co_xMo_yO and Ni_xCo_yO. Commonly used methods for the synthesis of binary and ternary metal oxides are solvothermal, microwave-assisted method and electrodeposition.

3.4.3 CHALCOGENIDES

Metal chalcogenides (MX; X = S, Se, Te) are compounds composed of electropositive element (metal) and chalcogenides (sulphur, selenium and tellurium). They have gained a great deal of interest due to their long stability and high power density. The supercapacitive performance of chalcogenide-based electrodes is influenced by factors such as chemical composition, crystalline phase and morphology. Most of them have a resemblance in their properties with pristine graphene. Not only binary, but ternary and higher order chalcogenides have also been successfully synthesized as electrode materials for supercapacitors. They can facilitate rich redox reactions and can accommodate broad range of electrolyte ions in the tuneable gap between the layers. Transition metal suphides such as nickel sulphides, copper sulphides, cobalt sulphides, $NiCo_2S_4$, $MnCo_2S_4$, Bi_2S_3, La_2S_3, WS_2, MoS_2, metal selenides such as $NiSe_2$, $CuSe$, $MoSe_2$ and cobalt selenides are some of the metal chalcogenides used as electrode materials for supercapacitors [18, 29, 30].

3.4.3.1 Nickel Sulphides

Nickel sulphides exist in several different forms such as NiS, NiS_2, Ni_3S_2, Ni_3S_4, Ni_7S_{10} and Ni_9S_8 and have different morphologies. Most of them are composed of more than one phase; hence synthesizing pure phase nickel sulphides is comparatively difficult. Ni_3S_2 nanostructures on nickel foam synthesized by Chou et al. through the potentiodynamic deposition method delivered a specific capacitance of 717 F/g at 2 A/g [31, 32].

3.4.3.2 Cobalt Sulphides

Researchers have identified different nanostructures of cobalt sulphides as efficient materials for supercapacitor electrodes. Synthesis of pure cobalt sulphides is complicated, since they easily undergo phase transformation. During the synthesis of cobalt sulphide, it is very difficult to control the reaction temperature as well as removing the impurities such as cobalt hydroxide and cobalt oxide. Some reported cobalt sulphides for supercapacitor electrodes are Co_3S_4, CoS, CoS_2 and Co_9S_8. Patil et al. synthesized reduced Graphene Oxide (rGO)-wrapped Co_3O_4 nanoflakes capable of delivering a high specific capacitance of 2314 F/g at 2 mV/s [33].

3.4.3.3 Copper Selenides

Copper selenides are inexpensive, possess high electrical conductivity and are capable of delivering better electrochemical properties. CuSe/Cu electrodes synthesized by Pazhamalai et al. as binder-free supercapacitor electrodes via the hydrothermal method delivered a specific capacitance of 1037.5 F/g at 0.25 mA/cm² [34].

3.4.4 CONDUCTING POLYMERS

Conducting polymers have been widely investigated as electrode material for super-capacitors due to their high electrical conductivity, large theoretical specific capacitance, low cost, ease of large-scale production and environmental friendliness. They store charge by the faradaic mechanism. Since participants of the charge storage process are polymeric chains, they are capable of giving high specific capacitance and high energy density. An effective way of increasing the specific energy density of conducting polymers is to incorporate a dopant. P-doping of a conducting polymer with an anion is possible during oxidation while n-doping with a cation can be achieved during reduction. Charging equations of these types of doping can be represented as follows [10]:

$$Cp \rightarrow Cp^{n+}\left(A^{-}\right)_{n} + ne^{-} \left(p-doping\right)$$

$$Cp + ne^{-} \rightarrow \left(C^{+}\right)_{n} Cp^{n-} \left(n-doping\right)$$

Conducting polymer-based supercapacitors can be of three types. If both the electrodes of the device are made of the same material, then they belong to type I. In type II devices, the two electrodes are made from two different p-doped conducting polymers. A type III device is composed of electrodes made of the same conducting polymers, where one is p-doped while the other is n-doped. Selection of a potential window is very crucial for the better operation of supercapacitors based on conducting polymers. Because they may either degrade (at more positive potential) or may be transformed into an insulating state (at more negative potential). The major obstacle faced by conducting polymer-based electrodes for practical applications is their poor cycling stability, which is due to the swelling and shrinkage during a continuous charge/discharge process. There are many ways that the researchers have been trying to improve the cycling life of conducting polymers, such as compositing with metal oxides/sulphides/hydroxides, altering the morphology and making the hybrid of the polymers. Most widely used conducting polymers include polyaniline (PANI), polypyrrole (PPy) and polythiophene (PTh) [10, 13, 18, 35].

3.4.4.1 Polyaniline (PANI)

PANI has a number of advantages, such as environmental stability and ease of synthesis when used as an electrode material for supercapacitors. The morphology of PANI nanostructures strongly influences their electrochemical properties. Therefore, selection of an appropriate high-efficiency method is crucial for the synthesis of PANI with suitable nanostructures. PANI has theoretical specific capacitance of 2000 F/g, but the experimental value is much lower than the theoretical value due to the small percentage contribution of PANI towards capacitance ability. A supercapacitor device constructed by using PANI nanofibres synthesized by Sivakumar et al. via interfacial polymerization delivered a specific capacitance of 554 Fg^{-1} at 1 A^{-1}. However, the cycling stability of pure PANI still couldn't meet the requirement for practical applications. Researchers have been trying to composite PANI with carbon materials/metal oxides to enhance its electrochemical performance [35].

3.4.5 Nanostructured Carbon

Nanostructured carbon materials have attracted a great deal of interest as a supercapacitor electrode material due to their high specific surface area, good electrical conductivity, fast electron transfer kinetics, corrosion resistance, environmental friendliness, high temperature tolerance and low fabrication cost. The charge storage in these materials is a result of surface adsorption of electrolyte ions at the electrode/electrolyte interface following an EDLC mechanism, which is discussed in Chapter 2; however, it is a common practice to incorporate these carbon-based materials into psudocapacitive electrodes to improve the conductivity and hence overall efficiency.

3.5 ELECTROLYTES

The electrolyte forms an integral part of any energy storage device apart from the cathode and anode. An electrolyte is any substance (salt or the solute) that releases ions when dissolved in a suitable solvent. It is the medium which helps the transfer of ions between the cathode and the anode. It behaves as a vehicle that only ions can ride on, not the electrons. Thus, electrons flow through the connecting wires carrying power to the external load while the movement of ions through the electrolyte completes the circuit. Both the solute and the solvent influence the behaviour of the electrolyte and its suitability with a particular electrode material. While the electrode materials determine the basic performance of a supercapacitor, the electrolyte and separator determine the safety of the system.

3.5.1 Classification of Electrolytes

Electrolytes can be broadly classified into two types: solid-/quasi-solid-state electrolytes and liquid electrolytes. Aqueous electrolytes, organic electrolytes and room temperature ionic liquids (RTILs) are the different types of liquid electrolytes. Aqueous electrolytes mainly fall into three categories, viz. alkaline (e.g., NaOH/KOH), acidic (e.g., H_2SO_4) and neutral (e.g., Na_2SO_4). In spite of advantages such as low cost, high ionic conductivity and environmental friendliness, narrow operating voltage ranges limit their large-scale applications.

In order to transcend the limitations of aqueous electrolytes, alternatives such as organic electrolytes and ionic liquids are being explored. Most of the commercially available supercapacitors use organic electrolytes such as those offering a wide operating voltage window, as large as 2.7 V. However, high cost, toxicity and flammability are some of the key factors that restrain the use of organic electrolytes as compared to their aqueous counterparts.

Properties such as high electrochemical stability, negligible volatility and non-flammability have made ionic liquids another potential candidate for supercapacitor electrolytes. Their high viscosity and high cost are the factors that hinder the possibility of their practical applications.

3.5.2 CRITERIA FOR SELECTION OF ELECTROLYTE

The choice of the electrolyte is a crucial factor that affects the overall performance of the supercapacitor device being fabricated. The main parameters that decide the suitability of an electrolyte for a specific device include ionic conductivity, stable operating voltage range, cost, safety, etc.

The electrolyte must possess high ionic conductivity so as to obtain maximum power density, as is evident from the relation $P = \dfrac{V^2}{4R}$, where R is the equivalent series resistance (ESR) and V is the operating potential. The electrolyte resistance is one of the major contributors to the ESR of the device, along with that offered by the current collector and separator. Moreover, the ionic size of the electrolyte ions should be compatible with the pore size of the electrode material so as to improve the intercalation/deintercalation process during the charging/discharging of the device. Hence, electrolytes with high ionic conductivity and suitable ionic radius should be selected to enhance the overall performance of the supercapacitor device. It is worthwhile to note that the electronic conductivity of the electrolyte must be negligible so as to force the free electrons to flow through the external load during discharging, delivering power.Since the energy density of a supercapacitor varies as the square of its operating voltage (V), as represented by the relation, $E = \dfrac{1}{2}CV^2$, emphasis must be laid on selecting components that provide a wide and stable potential window – defined as the voltage range in which the electrolyte is not oxidized or reduced. The stable operating potential window of a supercapacitor device largely depends on the nature of the solvent used in the preparation of the particular electrolyte. In the case of aqueous electrolytes (such as KOH, H_2SO_4, Na_2SO_4 and so on), the usable potential window is practically limited to around one volt as water undergoes thermodynamic decomposition at 1.23 V (vs Reversible Hydrogen Electrode (RHE)). As the applied potential goes beyond this limit, oxygen gas is generated at the positively charged anode and hydrogen gas is liberated at the negatively charged cathode as the result of the electrolysis of water, also known as water splitting. Oxygen and hydrogen gases released during this process form bubbles around the respective electrodes, causing surface hindrance and thus reducing the effective electrode area in contact with the electrolyte.

3.5.3 ADDITIVES IN ELECTROLYTES

One method to compensate the reduced energy density due to limited operating voltage is to introduce suitable redox active substances such as $K_3[Fe(CN)_6]$ into aqueous electrolytes. They contribute extra redox reactions in the system, resulting in enhanced specific capacitance and coulombic efficiency [36]. Redox couples like halides (I^-/I^{3-} and Br^-/Br^{3-}), vanadium complexes, copper salts, methylene blue, phenylenediamine, indigo carmine, quinones (Q) and hydroxyquinone (HQ) etc., are also being explored. The use of $Fe(CN)_6^{4-}/Fe(CN)_6^{3-}$ as a redox additive in the electrolytes for supercapacitors has been reported by many researchers. According to Nagaraju et al., introduction of a small quantity of highly soluble and electrochemically active hexacyanoferrate ions into aqueous KOH solution could considerably improve the specific capacitance of Ni_3Se_2-based superapacitor devices [37]. A nearly twofold improvement in the specific capacitance and energy density has been achieved in symmetrically activated carbon supercapacitors by the addition of KI into H_2SO_4 electrolyte [38]. Addition of Cu^{2+}/Cu^+ couple in aqueous electrolytes has resulted in a tenfold increase in the specific capacitance of supercapacitors based on porous carbon microspheres [39]. Gao et al. have reported a substantial improvement in the capacitance of KI-H_2SO_4 mixed electrolyte Â ($\approx$616 Fg^{-1}) compared to the H_2SO_4 electrolyte ($\approx$184 Fg^{-1}) [40]. The use of organic compounds with functional groups such as quinone, hydroquinone, amine, etc., also has been reported to have produced exotic results [41].

As no single electrolyte can have all desirable characteristics, optimization of desirable properties is the key to formulate efficient electrolytes compatible with different electrode materials. Extensive research is being carried out to explore new electrolytes to improve the overall performance of supercapacitors.

3.6 SEPARATORS

The separator is a thin insulating membrane with submicron-sized pores that separates the cathode from the anode internally. The main role of the separator is to prevent short-circuiting the cell, i.e., it acts as a physical barrier to avoid the chance of direct electrical contact between the cathode and the anode. By carefully choosing separators with a suitable pore size, ions can be selectively passed through it while preventing the flow of electrons. The desirable properties of a separator include high porosity to electrolyte ions, inertness to electrolyte, mechanical stability, flexibility, etc. Cellulose and its derivatives, synthetic resins like polypropylene (PP), polyethylene (PE), glass-fibre and PTFE are some of the commonly used separators.

3.7 CURRENT COLLECTORS

The role of current collectors is to conduct current between the electrodes and the external load. Materials with high electrical conductivity, mechanical strength, chemical stability and flexibility are suitable for use as current collectors.

Metals like copper, aluminium and nickel are commonly used as current collectors in most of the commercial supercapacitors due to their high electrical conductivity and low cost. In addition to this, indium-tin-oxide (ITO), stainless steel, gold, tita-nium, etc., are also used.

Nickel foam is one of the most popular among current collectors due to its porous structure, compatibility with aqueous electrolytes and comparatively low cost. Xing et al. have opined that the use of nickel foam as a current collector which often results in over-estimation of the supercapacitor performance, especially when the mass loading of the electrode is very low [42]

Metallic current collectors should be cleaned thoroughly before use to enhance their performance. The oxide layers on their surface may increase the resistance, thereby increasing the ESR of the cell. Rinsing with acid or dilute etchants is highly recommended to overcome this issue.

3.8 CONCLUSIONS

Supercapacitors are being looked upon as the ideal candidate for meeting the ever-increasing demand for sustainable, environmentally benign and efficient energy storage systems coupled with high energy density and power density. Despite the technological advancements in this field, the limited value of the energy density of supercapacitors compared to that of batteries and fuel cells still remains a challenge in their wider applications. Researchers continue their efforts to address this issue by engineering suitable materials for various components of supercapacitors – the electrode, the electrolyte, the current collector, separator, etc., which can improve the capacitance as well as operating potential window of the device. In addition to carbon-based electric double-layer systems and pseudo capacitance devices, nanostructured composite materials are also being explored to design better supercapacitive devices as they possess a high specific surface area, offer faster kinetics and provide more electroactive sites for faradaic energy storage. Design, development and optimiza-tion of efficient supercapacitor systems is vital for the improvement of energy storage capability and power quality, facilitating faster access to the stored energy systems.

REFERENCES

1. Yu, X., Lu, B., & Xu, Z. (2014). Super long-life supercapacitors based on the con-struction of nanohoneycomb-like strongly coupled $CoMoO_4$–3D graphene hybrid elec-trodes. *Advanced Materials*, 26(7), 1044–1051.
2. Wang, J. G., Liu, H., Liu, H., Hua, W., & Shao, M. (2018). Interfacial constructing flex-ible V_2O_5@ polypyrrole core–shell nanowire membrane with superior supercapacitive performance. *ACS Applied Materials & Interfaces*, 10 (22), 18816–18823.
3. Chen, J., Zhang, Y., Hou, X., Su, L., Fan, H., & Chou, K. C. (2016). Fabrication and characterization of ultra light SiC whiskers decorated by RuO_2 nanoparticles as hybrid supercapacitors. *RSC Advances*, 6(23), 19626–19631.
4. Liu, X., Zhao, J., Cao, Y., Li, W., Sun, Y., Lu, J., & Hu, J. (2015). Facile synthesis of 3D flower-like porous NiO architectures with an excellent capacitance performance. *RSC Advances*, 5(59), 47506–47510.
5. Hu, B., Qin, X., Asiri, A. M., Alamry, K. A., Al-Youbi, A. O., & Sun, X. (2013). Synthesis of porous tubular C/MoS_2 nanocomposites and their application as a novel electrode material for supercapacitors with excellent cycling stability. *Electrochimica Acta*, 100, 24–28.

6. Muzaffar, A., Ahamed, M. B., Deshmukh, K., & Thirumalai, J. (2019). A review on recent advances in hybrid supercapacitors: Design, fabrication and applications. *Renewable and Sustainable Energy Reviews*, 101, 123–145.

7. Yan, J., Khoo, E., Sumboja, A., & Lee, P. S. (2010). Facile coating of manganese oxide on tin oxide nanowires with high-performance capacitive behavior. *ACS Nano*, 4(7), 4247–4255.

8. Gopi, C. V. M., Vinodh, R., Sambasivam, S., Obaidat, I. M., & Kim, H. J. (2020). Recent progress of advanced energy storage materials for flexible and wearable supercapacitor: From design and development to applications. *Journal of Energy Storage*, 27, 101035.

9. Scibioh, M. A., & Viswanathan, B. (2020). *Materials for Supercapacitor Applications*. Elsevier.

10. Kar, K. K. (2020). *Handbook of Nanocomposite Supercapacitor Materials II*. Springer International Publishing.

11. Fleischmann, S., Mitchell, J. B., Wang, R., Zhan, C., Jiang, D. E., Presser, V., & Augustyn, V. (2020). Pseudocapacitance: from fundamental understanding to high power energy storage materials. *Chemical Reviews*, 120(14), 6738–6782.

12. Come, J., Augustyn, V., Kim, J. W., Rozier, P., Taberna, P. L., Gogotsi, P., & Simon, P. (2014). Electrochemical kinetics of nanostructured Nb_2O_5 electrodes. *Journal of the Electrochemical Society*, 161(5), A718.

13. Bard, A. J., Inzelt, G., & Scholz, F. (Eds.). (2008). *Electrochemical Dictionary*. Springer Science & Business Media.

14. Conway, B. E. (1993). Two-dimensional and quasi-two-dimensional isotherms for Li intercalation and UPD processes at surfaces. *Electrochimica Acta*, 38(9), 1249–1258.

15. Conway, B. E., & Angerstein-Kozlowska, H. (1981). The electrochemical study of multiple-state adsorption in monolayers. *Accounts of Chemical Research*, 14(2), 49–56.

16. Costentin, C., Porter, T. R., & Savéant, J. M. (2017). How do pseudocapacitors store energy? Theoretical analysis and experimental illustration. *ACS Applied Materials & Interfaces*, 9(10), 8649–8658.

17. Conway, B. E. (2013). *Electrochemical Supercapacitors: Scientific Fundamentals and Technological Applications*. Springer Science & Business Media.

18. Brousse, T., Bélanger, D., & Long, J. W. (2015). To be or not to be pseudocapacitive?. *Journal of the Electrochemical Society*, 162(5), A5185.

19. Augustyn, V., Simon, P., & Dunn, B. (2014). Pseudocapacitive oxide materials for high-rate electrochemical energy storage. *Energy & Environmental Science*, 7(5), 1597–1614.

20. Levi, M. D., & Aurbach, D. (1999). Frumkin intercalation isotherm: A tool for the description of lithium insertion into host materials: A review. *Electrochimica Acta*, 45(1–2), 167–185.

21. Liu, T., Pell, W. G., & Conway, B. E. (1997). Self-discharge and potential recovery phenomena at thermally and electrochemically prepared RuO2 supercapacitor electrodes. *Electrochimica Acta*, 42(23–24), 3541–3552.

22. Soin, N., Roy, S. S., Mitra, S. K., Thundat, T., & McLaughlin, J. A. (2012). Nanocrystalline ruthenium oxide dispersed Few Layered Graphene (FLG) nanoflakes as supercapacitor electrodes. *Journal of Materials Chemistry*, 22(30), 14944–14950.

23. Ali, G. A., Tan, L. L., Jose, R., Yusoff, M. M., & Chong, K. F. (2014). Electrochemical performance studies of MnO_2 nanoflowers recovered from spent battery. *Materials Research Bulletin*, 60, 5–9.

24. Toupin, M., Brousse, T., & Bélanger, D. (2004). Charge storage mechanism of MnO_2 electrode used in aqueous electrochemical capacitor. *Chemistry of Materials*, 16(16), 3184–3190.

25. Nam, K. W., & Kim, K. B. (2002). A study of the preparation of NiO_x electrode via electrochemical route for supercapacitor applications and their charge storage mechanism. *Journal of the Electrochemical Society*, 149(3), A346.

26. Ge, C., Hou, Z., He, B., Zeng, F., Cao, J., Liu, Y., & Kuang, Y. (2012). Three-dimensional flower-like nickel oxide supported on graphene sheets as electrode material for supercapacitors. *Journal of Sol-gel Science and Technology*, 63(1), 146–152.

27. Kate, R. S., Khalate, S. A., & Deokate, R. J. (2018). Overview of nanostructured metal oxides and pure nickel oxide (NiO) electrodes for supercapacitors: A review. *Journal of Alloys and Compounds*, 734, 89–111.

28. Rakhi, R. B., Nagaraju, D. H., Beaujuge, P., & Alshareef, H. N. (2016). Supercapacitors based on two dimensional VO_2 nanosheet electrodes in organic gel electrolyte. *Electrochimica Acta*, 220, 601–608.

29. Palchoudhury, S., Ramasamy, K., Gupta, R. K., & Gupta, A. (2019). Flexible supercapacitors: A materials perspective. *Frontiers in Materials*, 5, 83.

30. Shaikh, N.S., Ubale, S.B., Mane, V.J., Shaikh, J.S., Lokhande, V.C., Praserthdam, S., Lokhande, C.D., & Kanjanaboos, P. (2021). Novel electrodes for supercapacitor: Conducting polymers, metal oxides, chalcogenides, carbides, nitrides, MXenes, and their composites with graphene. *Journal of Alloys and Compounds*, 893, 161998.

31. Shaikh, N. S., Ubale, S. B., Mane, V. J., Shaikh, J. S., Lokhande, V. C., Praserthdam, S., & Kanjanaboos, P. (2022). Novel electrodes for supercapacitor: Conducting polymers, metal oxides, chalcogenides, carbides, nitrides, MXenes, and their composites with graphene. *Journal of Alloys and Compounds*, 893, 161998.

32. Theerthagiri, J., Karuppasamy, K., Durai, G., Rana, A. U. H. S., Arunachalam, P., Sangeetha, K., & Kim, H. S. (2018). Recent advances in metal chalcogenides (MX; X= S, Se) nanostructures for electrochemical supercapacitor applications: A brief review. *Nanomaterials*, 8(4), 256.

33. Patil, S. J., Kim, J. H., & Lee, D. W. (2017). Graphene-nanosheet wrapped cobalt sulphide as a binder free hybrid electrode for asymmetric solid-state supercapacitor. *Journal of Power Sources*, 342, 652–665.

34. Pazhamalai, P., Krishnamoorthy, K., & Kim, S. J. (2016). Hierarchical copper selenide nanoneedles grown on copper foil as a binder free electrode for supercapacitors. *International Journal of Hydrogen Energy*, 41(33), 14830–14835.

35. Meng, Q., Cai, K., Chen, Y., & Chen, L. (2017). Research progress on conducting polymer based supercapacitor electrode materials. *Nano Energy*, 36, 268–285.

36. Samanta, P., Ghosh, S., Murmu, N. C., & Kuila, T. (2021). Effect of redox additive in aqueous electrolyte on the high specific capacitance of cation incorporated MnCo2O4@ Ni (OH) 2 electrode materials for flexible symmetric supercapacitor. *Composites Part B: Engineering*, 215, 108755.

37. Nagaraju, G., Cha, S. M., Sekhar, S. C., & Yu, J. S. (2017). Metallic layered polyester fabric enabled nickel selenide nanostructures as highly conductive and binderless electrode with superior energy storage performance. *Advanced Energy Materials*, 7(4), 1601362.

38. Senthilkumar, S. T., Selvan, R. K., Lee, Y. S., & Melo, J. S. (2013). Electric double layer capacitor and its improved specific capacitance using redox additive electrolyte. *Journal of Materials Chemistry A*, 1(4), 1086–1095.

39. Mai, L. Q., Minhas-Khan, A., Tian, X., Hercule, K. M., Zhao, Y. L., Lin, X., & Xu, X. (2013). Synergistic interaction between redox-active electrolyte and binder-free functionalized carbon for ultrahigh supercapacitor performance. *Nature Communications*, 4(1), 1–7.

40. Gao, Z., Zhang, L., Chang, J., Wang, Z., Wu, D., Xu, F., & Jiang, K. (2018). Catalytic electrode-redox electrolyte supercapacitor system with enhanced capacitive performance. *Chemical Engineering Journal*, 335, 590–599.
41. Roldán, S., Blanco, C., Granda, M., Menéndez, R., & Santamaría, R. (2011). Towards a further generation of high-energy carbon-based capacitors by using redox-active electrolytes. *Angewandte Chemie International Edition*, 50(7), 1699–1701.
42. Xing, W., Qiao, S., Wu, X., Gao, X., Zhou, J., Zhuo, S., & Hulicova-Jurcakova, D. (2011). Exaggerated capacitance using electrochemically active nickel foam as current collector in electrochemical measurement. *Journal of Power Sources*, 196(8), 4123–4127.

4 Looking Deeper into Electrode Processes

*Sindhu Thalappan Manikkoth, Fabeena Jahan,
Deepthi Panoth, Anjali Paravannoor, and
Baiju Kizhakkekilikoodayil Vijayan*

CONTENTS

4.1 INTRODUCTION

Electrochemical capacitors are considered as efficient and alternative clean energy storage compared to batteries and fuel cells. They have found numerous applications in the field of electric vehicles, portable and wearable electronics, and other energy storage systems owing to their extraordinary characteristics such as high specific capacitance, long cycle life, enhanced power and energy density, fats charge/discharge, high-rate capability, and excellent reversibility. The different capacitor

DOI: 10.1201/9781003258384-4

architectures give rise to different types of supercapacitors such as double-layer capacitance, pseudocapacitors (PCs) and hybrid capacitors. The electrodes and electrolytes employed will decide the energy storage mechanism in supercapacitor devices. Generally, the charge/discharge systems at the electrode/electrolyte induce the energy storage capacity of supercapacitors. The range of charge storage for conventional capacitors displays a charge storage capacity within the micro to millifarad range, while supercapacitors present a charge storage range between 100 and 1000 F with maintaining specific power and low series resistance (ESR). Compared to batteries, supercapacitors also exhibit higher power density in several orders of magnitude and significant specific energy, making them a feasible substitute in energy storage systems. They also show reduced battery size, weight and cost, extended battery run time, provided energy storage and source balancing when used with energy harvesters, and battery life, enabled low/high-temperature operation, and minimized space requirements. They are also safe and have a greater charging/discharging power rate.

A number of electrode materials are involved in supercapacitor assembly including carbonaceous materials, transition metal oxides/hydroxides, and conducting polymers. Among the metal oxides, NiO, MnO_2, ZnO, and Co_3O_4 are better materials for pseudocapacitance where charge storage is accompanied by means of the fast faradaic redox reactions. The incorporation of nanoparticles in the metal oxides can also increase the capacitance and the rate of intercalation from the electrolyte to the electrode and vice versa. Similarly, the capacitance of metal oxides can also be enhanced with the addition of conductive carbon materials and conductive polymer materials. The morphology, pore distribution, structure, and the type of electrolyte employed also has a direct effect on supercapacitor performance.

This chapter discusses the effect of various electrode processes and electrode materials on supercapacitor performance. It also discusses the structure and porosity in the electrode performance and the methods for tuning or improving supercapacitor performance.

4.2 FUNDAMENTALS OF ELECTRIC DOUBLE-LAYER CAPACITANCE AND PSEUDOCAPACITANCE

The electrode material is considered a key component of supercapacitors which act as the site for energy storage. The major parameters of electrode materials that influence supercapacitor performance are surface area, porosity, pore size distribution, surface functionality, and electrical conductivity of the material [1]. In supercapacitors, charge storage mechanism occurs mainly based on two principles: electric double-layer capacitance, which is based on electrostatic interaction, and the other is pseudocapacitance which involves chemical reactions [2]. In electric double-layer capacitors (EDLCs) the capacitance is improved by selecting an electrode material with good electronic conductivity and a large surface area as the selection of electrode material plays a crucial role to enhance the electrochemical performance in supercapacitors. The energy density of supercapacitors is usually very low compared to batteries. In terms of energy density there still exists a wide gap between supercapacitors and batteries. Effective methods are being tested to improve the existing value by enhancing the active surface area of electrode materials in EDLCs or by enhancing the operation voltage window with the use of organic electrolytes. Researchers have been putting great effort into developing larger surface area materials, as with a larger active surface area the number of ions stored by forming the Helmholtz double layer increases which result in higher energy density. Thus, carbon-based nanostructured materials with excellent electronic conductivity and higher specific surface area are usually preferred as electrode materials for EDLCs which include activated carbon, graphene, carbon nanotubes (CNTs), etc. One of the major limitations of EDLC-based supercapacitors is their lower energy density compared to that of lithium-ion (Li-ion) batteries, even though cyclic stability is substantially better [3]. However, PCs are capable of storing charge by means of electric double-layer formation and also via reversible redox reactions which enable fast insertion of electrolytes on the electrode surface. In supercapacitors, with pseudocapacitance, the energy density can be increased by maintaining their cyclic stability (Figure 4.1).

Conway identified that the PCs show three different mechanisms which include (1) underpotential deposition, (2) redox pseudocapacitance, and (3) intercalation pseudocapacitance [4]. Underpotential deposition occurs when an adsorbed monolayer is formed on the metal electrode surface with an increased redox potential by the cations present in the electrolyte. The conventional example of underpotential deposition is lead deposited on the surface of the gold electrode at a specified potential [5]. Redox pseudocapacitance occurs on the surface or subsurface of the electrodes with concomitant faradaic charge transfer between the ions present in the electrolyte and the electrode [6]. The common classic examples of redox pseudocapacitive materials are metal oxides such as MnO_2, NiO, RuO_2, and Co_3O_4 and also conducting polymers such as polyaniline and polypyrrole. Intercalation pseudocapacitance consists of the intercalation of ions into the bulk layer or the tunnels of the redox-active material along with a faradaic charge transfer and even after the reaction there

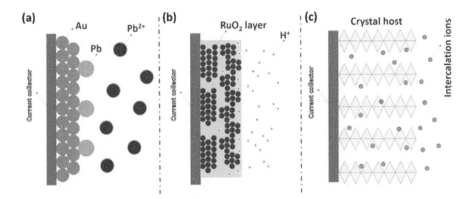

FIGURE 4.1 Schematic representation of different mechanisms, (a) underpotential deposition, (b) faradaic redox pseudocapacitance, and (c) intercalation pseudocapacitance. Reproduced from Liu et al. [3].

is no crystallographic phase change observed. Thus, in the electrochemical reaction, it retains structure stability [7].

4.2.1 MATERIALS WITH PSEUDOCAPACITIVE BEHAVIOR

The pseudocapacitive materials exhibit Cyclic Voltammetry (CV) and Galvanostatic Charge Discharge (GCD) profiles similar to that of EDLC materials showing linear dependence of charge storage with a potential window. This implies that they can store charge via both surface faradaic and double-layer mechanisms. In pseudocapacitive materials, the charge storage can occur in two ways which involve, (a) a surface redox reaction or (b) an intercalation type reaction [8]. Electrode materials which exhibit pseudocapacitive behavior can be categorized into two types: one is intrinsic PC materials and the other is extrinsic PC materials [9]. Intrinsic PC materials are capable of exhibiting capacitive charge storage behavior for a vast range of particle sizes and morphologies [10]. Commonly known intrinsic PC materials are RuO_2, MnO_2, Nb_2O_5, hydrogen titanate, and $TiO_2(B)$. Extrinsic PC materials exhibit pseudocapacitive behavior only when a substantial fraction of Li^+ storage sites are available on the electrode surface for a battery material engineered at the nanoscale. Common examples of extrinsic PC materials are V_2O_5, MoO_2, and Li_2CoO_2 [11]. At the nanoscale, in extrinsic PC materials on the surface and near-surface region speedy surface-dominated redox reactions occur that obey surface-confined electro-kinetics instead of diffusion-controlled kinetics. Unlike intrinsic PC materials, in bulk state, extrinsic materials do not display pseudocapacitive phase transformation and is suppressed by enhancing the surface area via nanostructuring which in turn results in reducing the diffusion distance and thus leads to increased high-rate behavior. There are specific features used to characterize and analyze the pseudocapacitive behavior of materials based on their response to: (i) a voltage sweep, in CV, (ii) constant current, in GCD, and (iii) alternating current, in EIS. And by carefully analyzing their response we determine whether the material exhibits pseudocapacitive behavior.

4.2.1.1 Intrinsic Pseudocapacitor Materials

Intrinsic PC materials are materials that deliver capacitance behavior across a wide range of particle sizes and dimensions. The fundamental candidates explored as intrinsic PC materials are transition metal oxides like RuO_2, MnO_2, etc., and conductive polymers [8]. Hydrous RuO_2 is considered as the first material to exhibit pseudocapacitive behavior owing to its rapid proton and electron-conducting properties [12]. RuO_2 is one of the most commonly used intrinsic PC materials. Though they exhibited a charge storage process, their cyclic voltammogram displayed a rectangular shape. $RuO_2.5H_2O$ displays various unique features that enable fast faradaic reactions with enhanced capacitance: (i) Ru^{2+} cations redox behavior, (ii) faster electron transport due to the high metallic conductivity of RuO_2, (iii) structural water presence helps in ion diffusion with larger "inner surface", and (iv) larger surface area reduces the diffusion distances. Despite all these features, the high cost and scarcity of ruthenium limit its application for large scale production. Later studies reported improving the capacitance with porous structures, nanoscale architectures, and by identifying the significance of structural water in $RuO_2.nH_2O$ (where n = 0.5). A capacitance value of 720 F/g was obtained with the traditional slurry method for a charge/discharge time of 8 minutes, which is about 50% of the theoretical capacitance of $RuO_2.nH_2O$ [13]. The capacitance value obtained for 10 wt% RuO_2 dispersed on 90 wt% activated carbon was 1340 F/g for a charge/discharge time of 1 minute which is far closer to the theoretical capacitance value [14]. The effective exposure of $RuO_2.nH_2O$ with the electrolyte resulted in enhanced capacitance and rate. In this type of electrode architecture, the major limitation is its very low areal capacitance values as the total mass loading of $RuO_2.nH_2O$ per footprint area is small.

MnO_2 is the next most commonly used intrinsic PC material. Lee and Goodenough studied the pseudocapacitive behavior of $MnO_2.nH_2O$ in 1999, where they observed a cyclic voltammogram of rectangular shape with a capacitance of 200 F/g in KCl aqueous electrolyte. The pseudocapacitive charge storage in MnO_2 is due to the rapid faradaic reactions and the redox reaction of Mn between +3 and +4 oxidation states at its surface or into the bulk. MnO_2 displays electrochemical properties with the characteristic of intrinsic pseudocapacitance. In manganese oxides their pseudocapacitive properties largely rely on their crystallographic structures and crystallinity [15]. Compared to RuO_2, MnO_2 is easily available, abundant, and its low cost makes it a better candidate for electrodes. The electronic conductivity of MnO_2 ranges from 10^{-7} to 10^{-3} S cm^{-1} [16]. As charge storage in MnO_2 takes place inside the thin layer of the surface and it gives rise to capacitance values ranging between 200 and 250 F g^{-1}, which is considerably lower compared to the theoretical capacitance value of thick composite electrodes. In MnO_2 nanostructuring is found to be an effective method for enhancing their specific capacitance as it is possible to access most of the MnO_2 storage sites and thus MnO_2 ultra-thin films can exhibit specific capacitance greater than 1000 F g^{-1} [17, 18]. Different strategies are being developed to enhance the electrochemical properties of MnO_2-based electrodes for supercapacitors. Brousse et al. studied the relation between specific capacitance and surface area for amorphous and crystalline MnO_2 compounds. The capacitance value of crystalline material was about 250 F g^{-1} which was slightly higher compared to the capacitance value of amorphous materials (~160 F g^{-1}). Here it was found that lower surface area resulted in higher capacitance values as crystalline MnO_2 possesses additional capacitance from bulk ion intercalation [19]. The major limitation in using MnO_2 electrodes is their poor electronic conductivity and lower cyclic stability, and thus various diverse approaches need to be developed to enhance its overall performance.

4.2.1.2 Extrinsic Pseudocapacitor Materials

Extrinsic PC materials are those exhibiting battery-like behavior in its bulk phase with strong redox peaks and plateau in their GCD profiles, though, in its nanophase it reveals pseudocapacitive behavior. As nanostructuring enhances the specific surface area of the electrodes and consequently the interfacial contact area between the electrode and electrolyte is increased with more reactive sites. This in turn reduces the ion-diffusion path and results in dispersion in the redox site energy [6]. The typical extrinsic PC materials in which PC behavior emerges at the nanoscale are V_2O_5, MoO_2, and Li_2CoO_2.

Vanadium oxides are an attractive candidate for electrochemical energy storage due to higher values of oxidation state which enables storage of more than one electron per unit (+2, +3, +4, and +5 are electrochemically available) and their capability to form layered structures. The pseudocapacitance behavior of high surface area V_2O_5 is heavily reliant on the exposure of active sites to the electrolyte. Vanadium oxides are an attractive candidate for electrochemical energy storage due to higher oxidation state values which enable storage of more than one electron per unit (+2, +3, +4, and +5 are electrochemically available) and their capability of forming layered structures. The pseudocapacitance behavior of high-surface-area V_2O_5 is heavily reliant on the exposure of active sites to the electrolyte [20]. The pseudocapacitive behavior of both amorphous and nanocrystalline V_2O_5 can be confirmed from the broad, featureless cyclic voltammogram and by the sloping charge/discharge curves obtained at the time of galvanostatic cycling. A particular class of materials was synthesized from V_2O_5 gel using a solvent removal technique into xerogels, aerogel, and ambigels. V_2O_5 xerogels structures possess V_2O_5 bilayers spaced apart from each other with a van der Waals gap of 12 Å that accommodates water molecules. The structure of ambigels and aerogels is identical to that of xerogels with a larger van der Waals gap ranging from 10–14 Å, as these larger gaps enable the insertion of a wide variety of cations [21]. Aerogel electrodes synthesized via the traditional composite electrode technique show a notable intercalation peak in their CV whereas V_2O_5 aerogels synthesized using the "sticky carbon" method display perfectly capacitive CV. The V_2O_5 aerogels exhibited a maximum capacitance of 1300 F g^{-1} due to the exposure of all active areas to the electrolyte [22].

MoO_2 is another common example of extrinsic pseudocapacitive material as it belongs to the space group P21/c, that is made from a tunnel structure framework which possesses a rutile-type structure with MoO_6 octahedra linkages. Along the a-axis the MoO_6 octahedra linkages form 1D channels which are capable of insertion and extraction of lithium ions reversibly. To attain higher capacity, a MoO_2 electrode utilizes both conversion reaction and the intercalation process. During the conversion reaction, the large volume expansion results in poor reversibility and slow reaction whereas the one-electron lithium intercalation is found to be favorable for pseudocapacitive charge storage mechanism. In contrast, the material experiences phase transformation during charging and discharging in the one-electron intercalation process which leads to poor rate capability. The phase transitions result in reduced conductivity. These limitations can be addressed by nanostructuring MoO_2. Kim et al. reported that MoO_2 is an exemplary material at nanoscale as the phase transformations are inhibited and in addition to that, nanostructure morphology offers short ion diffusion path lengths. Nanoscale architecture of MoO_2 displays superior energy storage kinetics than its bulk material [23].

4.2.2 INTERCALATION PSEUDOCAPACITANCE

Intercalation pseudocapacitance is a new form of charge storage mechanism which relies on the cation (Na^+, H^+, K^+, and Li^+) intercalation/deintercalation in the bulk of the active materials, however it's not restricted by the cation diffusion within the crystalline framework of active materials. The intercalation PC electrodes exhibits higher rate capability than the battery-type electrodes as there are no crystallographic phase transformations in the intercalation electrode materials. Nevertheless, the kinetics of intercalation pseudocapacitance is similar to that of the conventional pseudocapacitance, but its electrode profile is identical to that of the battery-type electrode where the charge-storage mechanism happens in too narrow a potential window [24]. Most intercalation pseudocapacitive materials exhibit hardly any dependence on the size of the particles and their morphology. The typical materials that exhibit intercalation pseudocapacitance are TiO_2-B, perovskite oxides, and T-Nb_2O_5. But in a few intercalation pseudocapacitive materials such as V_2O_5, $LiCoO_2$, and MoO_3, this behavior is exhibited only when the particle size is in nanoscale as the ion diffusion distance decreases in nanoparticles and nanostructuring helps in the suppression of phase transformation at the time of intercalation process which reduces the energy barrier. Augustyn et al. studied the intercalation pseudocapacitance behavior in T-Nb_2O_5 where it exhibited exceptionally high-rate capability, as its crystal structure offers faster ionic transport. Within short periods of time high levels of charge storage were possible as it showed two-dimensional transport pathways and minor structural change with intercalation [25]. In $LiCoO_2$ the intercalation pseudocapacitance behavior became more prominent with decreasing size and the capacitance enhanced when the particle size was reduced to nanoscale (not less than 15 nm) [26].

4.2.2.1 Cation Intercalation Pseudocapacitance

The major rule for designing cation intercalation pseudocapacitance in electrode materials is that it should possess a crystal structure with 2D rapid ion diffusion pathways and ultra-structural stability to suppress phase transformations during the intercalation process. Cations such as Na^+, Li^+, Mg^{2+}, K^+, and Al^{3+} exhibit intercalation pseudocapacitance, delivering capacitance more than 300 F cm^{-2}. The ionization energy and electronegativity reduce as the ionic radii of intercalated cations increase. To overcome the limitation of cycling stability and rate capability Zhao et al. intercalated appropriate alkali metal ions into the allowed layered structures [27]. Augustyn et al. demonstrated the pseudocapacitive mechanism of nanocrystalline Nb_2O_5 films in organic electrolytes, which exhibited a storage capacity of ~130 mAh g^{-1} [25]. The charge storage mechanism in Nb_2O_5 can be described as:

$$Nb_2O_5 + xLi^+ + xe^{-1} \leftrightarrow Li_xNb_2O_5$$

Where the maximum capacity $x = 2$. In Nb_2O_5 the intercalation pseudocapacitive behavior can be attributed to the surface redox reactions along with rapid transport of Li^+ ions through the crystal structure, which suppresses the phase change during the electrochemical reaction [3]. The crystal structure of T-Nb_2O_5 possesses unique open tunnels given by mainly vacant octahedral sites between (001) planes throughout the a-b plane, which enables the rapid transport of Li^+ structures. Titanium oxides and their compounds have been extensively explored as an electrode material with intercalation pseudocapacitance, as a charge storage mechanism, and their storage ability rely on the crystal structure. Li et al. found that the Li$^+$ ions intercalated and deintercalated into the TiO$_2$ lattice freely and reversibly. The intercalation mechanism in TiO$_2$ can be summarized as:

$$Ti_2 + x\left(Li^+ + e^{-1}\right) \leftrightarrow Li_xTiO_2$$

Wang et al. demonstrated that in TiO$_2$ intercalation pseudocapacitance can be enhanced by reducing the particle size. In addition to this at nanoscale these particles displayed a faster charging rate compared to larger particles [28]. Various approaches to enhance the intercalation capacitance of TiO$_2$ electrodes have been studied widely. Dopants were introduced into the oxide lattice structure, composites with carbon were prepared to address the issue of poor ionic conductivity of pristine TiO$_2$.

4.2.2.2 Anion Intercalation Pseudocapacitance

Oxygen ion-induced intercalation pseudocapacitance is gaining wide acceptance currently. In one charge/discharge cycle O^{2-} is capable of storing twice the charge with intercalation pseudocapacitance compared to the conventional Li$^+$ intercalation. Thus, compared to monovalent-ion intercalation oxygen ion intercalation is presumed to possess higher energy storage capacity. In 1990, Kudo first demonstrated the typical anion intercalation materials are perovskite oxides in aqueous alkaline solution [29]. The typical structural formula of perovskite oxides is ABO$_3$, where A and B cations possess a total positive valence of 6, which is equal to the total negative valence of oxygen anions. In the oxygen ion intercalation mechanism, initially from KOH electrolyte to active electrode surface, oxygen diffusion occurs in the form of hydroxide anions. Thereafter, with the diffusion of oxygen of hydroxide anions into the lattice with the oxidation of B elements, the oxygen vacancy is intercalated. In the meantime, in the electrolyte the proton is transferred to the hydroxide anion and water is obtained as the yield product. Finally, the oxygen intercalates via diffusion of B site elements into the perovskite lattice and surface of the lattice. Mainly there are three factors that determine the oxygen ion intercalation pseudocapactive behavior in perovskites which includes oxygen vacancy concentration, phase stability, and electrical stability. Thus, higher oxygen vacancy leads to better electrochemical performance of perovskite electrodes. Mefford et al. demonstrated oxygen anion-based intercalation pseudocapacitance and oxygen intercalation for fast energy storage in the perovskite LaMnO$_3$ with aqueous KOH electrolyte [30]. CaMnO$_3$ perovskite was reduced in 7%H$_2$/Ar gas atmosphere by Forslund et al. to obtain an electrode material with higher oxygen vacancies [31].

4.2.3 Carbon-Based Electrode Materials

Carbon electrodes are the most common EDLC materials where the capacitance is fully contributed from the electrostatic charge accumulated at the electrode/electrolyte interface and thus their electrochemical features depend crucially on the surface area of the electrodes [32]. Activated carbon is a carbonaceous material derived from charcoal, characterized by a high porous structure and large specific surface area.

The meso- and macropores permit the diffusion of charged species whereas micropores support the accumulation of charged species and thus enhance the specific capacitance [33]. Supercapacitors with higher energy density often exhibit relatively lower power density. One of the highly challenging and crucial tasks is to attain a compromise between energy density and power density. Thus, the design and control of the microstructure and porosity of activated carbon-based electrodes are considered as one of the best approaches to tackle this issue. Wang et al. developed two different porous structured electrode materials based on activated carbon with high micro- and mesoporous texture. The results revealed that microporous textures with high specific surface area exhibited higher specific capacitance than the mesoporous texture electrode. In order to enhance the energy density of supercapacitors by maintaining their power density and cycle life, designing activated carbons with narrow pore size distribution, with short pore length, and continuously interconnected pore structure would be an effective method.

4.3 EFFECT OF STRUCTURE AND POROSITY ON ELECTROCHEMICAL PERFORMANCE OF SUPERCAPACITORS

Electrode materials, electrolytes, current collectors, and separators play a major role in determining the electrical properties of a supercapacitor. Among these, electrode features such as morphologies and compositions, are largely responsible for the electrochemical performance of supercapacitors. Supercapacitors are of three types: EDLCs, PCs, and hybrid capacitors can be identified based on the cell configuration or storage mechanism. In comparison to electrostatic capacitors, EDLCs are primarily based on high specific-surface-area nanoporous materials as active electrode materials, resulting in a large capacitance and they can be controlled with respect to the area between the electrode and electrolyte, while PCs are made out of conducting polymer or metal oxide-based electrodes, as well as functionalized porous carbons. When compared to EDLCs, PCs possess high specific capacitance values and PC electrode materials have a short lifecycle [34].

The criteria for designing a high-performance supercapacitor electrode include high specific capacitance, large rate capability, high energy and power density, high cycle stability, as well as flow capacity. In addition, the toxicity and cost of the active materials used in an electrode design should be taken into account as well. The aforementioned properties are acquired by electrodes possessing the following features. (a) High surface area per gram: since charges are stored on the surface of the supercapacitor electrodes, an electrode with a higher surface area leads to an improved specific capacitance. (b) Controlled porosity: porous structure of electrode greatly influences the specific capacitance and the rate capability value. (c) Electroactive sites: they enable pseudocapacitance. (d) High electronic and ionic conductivity: specific capacitance, power density, and rate capability depend considerably on both electronic and ionic conductivity by maintaining the rectangular nature of cyclic voltammetry curve and symmetricity of galvanostatic charging/discharging curves. They also reduce the specific capacitance losses as scan rates/current densities are increased. A distinctive approach to enhance the electronic conductivity include binder-free electrode design and nanostructured current collector design to provide efficient electron pathways for charge transport. To increase the ionic conductivity, precise control of pore size and prudent design strategies are used. (e) Mechanical and chemical stability: the cycle stability is greatly influenced by the mechanical and chemical stability of electrode materials during cycling. Sensible electrode surface protection may substantially boost the cycle stability [35].

Since the electrode material is the most important component of a supercapacitor and controls the electrochemical performance, the development of new advanced electrode materials with rational nanostructured patterns has become a research hotspot in recent years. Many studies on supercapacitors have been reported that use advanced electrodes with enhanced surface characteristics and porous texture possessing high specific capacitance values with outstanding stability and cyclability, but only a few studies were conducted that specified the relationship between the structure and electrochemical performance of a supercapacitor. This section reviews the advances in nanostructured supercapacitor electrode materials, with emphasis on the understanding of the relationship between the structural as well as the porous properties of the electrode with electrochemical performance.

4.3.1 EFFECT OF POROSITY ON ELECTROCHEMICAL PERFORMANCE

For supercapacitor applications, electrodes with specific pore size, surface area, size distribution, etc., offer a major contribution to their performance. For an improved electrochemical reaction, the kinetics of ionic diffusion within the medium, as well as the electrolyte access, were greatly controlled by the porosity of a porous electrode. The kinetics of the electrochemical accessibility depends on the electrode surface, which in turn depends on the pores of different sizes. In order to design the porous electrodes for the different power and energy requirements, it must be taken accepted that not all pores in the matrix of a porous electrode are electrochemically accessible at the same time. In addition to the kinetic factors of the pore size distribution, the electrical conductivity is another limiting factor for the power density. The electrical conductivity or the ohmic resistance of porous materials is closely related to their morphologies. In general, the higher the surface, the smaller the particle size and the worse the conductivity should be. The mobility of the ions within the porous matrix is responsible for the speed of electrochemical accessibility and obviously, the movement of the ions in smaller pores is more difficult than in larger pores. Since the ionic diffusion in larger-sized pores is faster than that in smaller pores, they will deliver high energy at higher speed thus can be used for high-power applications [36].

According to the International Union of Pure and Applied Chemistry, pores were categorized into three types: macropores (more than 50 nm), mesopores (between 2 and 50 nm), and micropores (less than 2 nm). Proper mesopores could reduce ion transport resistance inside active materials during the charge/discharge process, providing a favorable pathway for quick movement of electrolyte ions due to their easy accessibility, thereby improving power output and rate capability of electrode materials at high current density. For capacitance retention and efficiency, it is essential to have a large average pore size and high mesopore porosity. But for porous carbon materials specific surface area is found to decrease with increasing pore size, implying that there should be a balanced compromise between energy density and power density in supercapacitors [37]. At low current densities, micropores can deposit more electrolyte ions on their surface, which promotes in-depth interfacial contacts and increases the EDLC, hence raising the achievable greatest specific capacitance and energy density. The major characteristics that dictate the high performance of double-layer capacitors are a large specific surface area, a hierarchically structured pore size distribution, and a high proportion of micropores. Macropores, as a characteristic microstructure with open gaps, can not only provide more sites for quick intercalation and deintercalation of active species, but also operate as ion-buffering reservoirs to reduce electrolyte diffusion distances and hence favor high capacitance output at high current densities. Furthermore, they can improve the loading and dispersion of other active components used in the production of composites [38]. As a result, it is vital to explain how to build acceptable porous designs. The electrolyte access, ion diffusion, electron transport into porous structures, and availability of active sites for electrochemical reactions are primarily determined by variables like pore size, pore geometry and distribution, surface curvature and chemistry, electrical conductivity, and the electrode/electrolyte interface. The electrical conductivity of active materials has a direct effect on electron transport, which affects the charge/discharge efficiency and power density of electrodes made from them. For inorganic material, increasing the preparation temperature improves crystallinity and thus conductivity. Too high a temperature, on the other hand, may eliminate existing organic groups on the surface of active materials. This would not only diminish electrolyte contact at the electrode/electrolyte interface, but it would also reduce organic groups' pseudocapacitance contribution [39]. Surface wettability of a porous electrode towards a given electrolyte is also an important factor to facilitate the adsorption and transportation of ions from the electrolyte which can be obtained by the introduction of surface functional groups, but this may decrease the electrical conductivity as they might inhibit the transport of electrons within the electrode [37]. Chemical modification and doping can improve electrolyte access and pseudocapacitance output by increasing the concentration of organic groups on surfaces [40]. Excessive organic groups, on the other hand, may cause interface electric resistance and so reduce capacitance output. It is clear that increased porosity promotes electrolyte infiltration and ion diffusion, resulting in higher capacitance output. However, excessive porosity may result in the breakdown of the electron transport network in active materials, lowering capacitance performance. As a result, the hierarchical porous structure with sufficient high porosity, electrical conductivity, an appropriate amount of surface organic groups, and coexisting macro-, meso-, and micropores should be the ideal construction for existing supercapacitor electrode materials to have enhanced performance of the supercapacitor that should be well matched with an electrolyte, and to have better electrochemical reaction activity. Jung et al. used electrochemical exfoliation process for the fabrication of a graphene electrode with meso- and macroporous structured aerogels with high specific surface area by optimizing certain factors like electrolyte content and freezing temperature in order to have a supercapacitor with extremely low densities and superior electrical properties. Resulting supercapacitor exhibited a specific capacitance of 325 Fg^{-1} at a current density of 1 A/g and an energy density of 45 Wh/kg in aqueous electrolyte of 0.5 M H_2SO_4 [41].

4.3.2 EFFECT OF STRUCTURE ON ELECTROCHEMICAL PERFORMANCE

The nanostructuring of electrode materials is a feasible method to considerably improve the surface area of the electrodes and hence the specific capacity. Typically, nanomaterials can be classified into zero dimensional (0D), one dimensional (1D), two dimensional (2D), and three dimensional (3D) categories. Particles which are more or less spherical in shape with three dimensions that are constrained on the nanoscale like fullerenes, quantum dots, nanoonions, nanoparticles, etc., are considered as 0D. There are three subcategories of 0D nanostructures used as electrodes in supercapacitors: solid, hollow, and core–shell 0D nanostructures [42]. In 2011, Chen and co-workers created nanoporous gold/MnO_2 hybrid supercapacitor electrodes to improve the electrical conductivity of MnO_2 and the resulting supercapacitor electrode demonstrated a high specific capacitance of 1145 F/g at 50 mV/s, which is due to nanoporous gold, that allows for easy and efficient access to electrons and ions [43]. Because of their unique features, including low density, high surface-to-volume ratio, and shorter paths for mass and charge transport, hollow 0D nanostructures have been identified as attractive options for electrode design [44]. Because of their amazing advantage in regulating the size, form, and structure of the products, hard templating methods have been widely employed for the synthesis of hollow 0D nanostrucures. Yang and co-workers in 2011 reported the fabrication of hollow carbon nanosphere with a large surface area, large bimodal mesopores, and large pore volumes generated by a silica sphere-assisted hard templating approach. A specific capacitance of 251 F/g at 50 mV/s can be achieved with this configuration [45].

Nanostructures with 1D are interesting due to their dimensionality dependence on their functional properties. While keeping the benefits of 0D nanostructures in the two nanoscale dimensions, the longitudinal axis of 1D nanostructures offer effective transport of both electrons and ions. They can be nanotubes, nanofibers/nanowires, nanorods/nanopillars, nanoribbons, and nanobelts. Nanowires have been explored extensively as an electrode material because of their high surface area-to-charge ratio for storing charges and efficient channel for transporting them [42]. Wei and colleagues described a simple one-step template-free method for making polyaniline nanowire arrays. The resulting polyaniline nanowire arrays at a current density of 1 A/g showed 950 F/g specific capacitance and 16% loss after 500 cycles [46]. Tubular nanostructures often offer more surface area with less mass usage than nanowires, resulting in more gravimetric specific capacitance. Because of their moderate to high surface area, porous structure, superior electronic conductivity, and excellent mechanical and thermal stability, CNTs have attracted a lot of attention as supercapacitor electrode materials. In 2010 Kim and others employed a water-assisted chemical vaporization deposition method to manufacture vertically aligned CNTs directly on conductive carbon sheets using an Al/Fe catalyst and shows a specific capacitance of roughly 200 F/g at a current density of 20 A/g [47].

Typically, materials with a thickness of a few atomic layers and with the other two dimensions beyond the nanometric size range are considered as 2D materials. Graphene and many other layered van der Waals solids like metal oxides and hydroxides and transition metal dichalcogenides and/or carbides and/or nitrides, etc., fall under the 2D category. The emergence of 2D materials, considering the relevant parameters like diffusion ion length or electrolyte contact area, as well as the fact that they only consider surface area for active material utilization in the performance of electrochemical supercapacitors, has attracted considerable attention for the hybrid 2D configurations in electrode material design. Due to its chemical stability as well as its unique intrinsic electrical, mechanical, and thermal properties, graphene is currently the most common material used for 2D material-based electrodes in energy storage devices. From various reports, reduced graphene oxide (GO) seems to be the widely used highly conductive electrode material and the mechanism employed for GO reduction become significant in the development of 2D electrode materials for supercapacitors. Ruoff and colleagues synthesized chemically modified graphene by the hydrazine hydrate reduction of suspended GO in water. In aqueous and organic electrolytes, the resultant sheets had a surface area of 705 m^2/g and yielded a specific capacitance of 135 and 99 F/g, respectively [48].

Materials with three dimensions beyond the nanometric size range but still preserving the advantages of nano size effect are regarded as 3D materials. For effective supercapacitor electrodes, 3D porous architectures of active materials provide a wide surface area, well-defined paths to electrolyte access, and mechanical stability. 3D electrodes are often made by organizing active materials into 3D nanostructures or utilizing metal foam as templates. Metal foams and carbonaceous material like mesoporous carbon and graphene aerogel fall under the 3D category. Cao Gaoping and co-authors found that developing porous carbon with suitable micropore size and adequate proportionate of mesopores is a more effective strategy for achieving high power density and high energy density for porous carbon-based supercapacitors. The carbon materials with mesopores and macropores boost energy density and power density of the EDLC [49, 50]. Metal foams like a Mn/MnO_2 core–shell with 3D macroporous structures were synthesized from ordered polystyrene sphere templates. Porous architectures of core–shell contribute several favorable features to effective supercapacitors, including a higher surface area, strong electrical conductivity attributable to Mn, and high capacitance of MnO_2. At 500 mV/s, the electrode's optimal specific capacitance is found to be 996 F/g, keeping 83% of the capacitance at 5 mV/s [51]. Recently Mu and coworkers carbonized waste polyethylene terephthalate to 3D porous carbon nanosheets (PCS) which were then hybridized with MnO_2 nanoflakes to generate PCS-MnO_2 composites. The resulting composite shows a specific capacitance of 210.5 F/g and a capacitance of 0.33 F/m^2 and also demonstrated an outstanding cycle stability with 90.1% of retention of capacitance [46].

Thus, we can conclude that the combination of various dimensionalities to produce a composite combining the properties of both EDLCs and PC electrodes will be beneficial for an enhanced supercapacitor performance with high conductivity, surface area, mechanical and performance stability, and capacitance.

4.4 TUNING THE PERFORMANCE OF SUPERCAPACITORS BY UNDERSTANDING THE CONCEPTS

In past decades, batteries had been considered as the most preferred energy storage systems for portable electronic devices owing to their great ability to store a large amount of energy, and the ability is called energy density. The higher energy density achieved with batteries astonished people and prevented them from exploring new energy storage systems for a long time. However, the long time taken for charging and discharging i.e., reduced power density is always a headache for battery researchers. The most demanding Li-ion batteries also can't overcome this issue even with the support of novel materials and technologies but resulted in increased weight and size, high cost, reduced energy, and cycle life. A miracle device architecture is difficult to find. Researchers tended to develop more advanced energy storage systems and it led to the fabrication of supercapacitors. Thus, supercapacitors have gained the attention of both academia and industry as an efficient alternative to conventional batteries which showed extraordinary features such as high specific capacitance from 1 mF to 10,000 F, Fast charge/discharge from milliseconds to seconds, excellent cycle life with a span of 5000 to 50,000 hours, environmentally friendly, low cost, wide operation potential and temperature, and the most important one, the high-power density from 0.01 to 10 kW/kg and energy density from 0.05 to 10 Wh/kg [52]. They also reduced battery size, weight, and cost, provided energy storage and source balancing when used with energy harvesters, extended battery run time and battery life, provided peak power and backup power, improved load balancing when used in parallel with a battery, lowered RF noise by eliminating DC/DC-enabled low/high temperature operation, cut pulse current noise, and minimized space requirements.

Despite the excellent features, supercapacitors show some restrictions to commercialization. The important issue is the relatively high cost per watt-hour of potential energy storage. Usage of high-cost and toxic electrode materials such as high-surface-area carbon materials and metal oxide-based electrodes also hinder large-scale production. The lower energy density with low-cost materials is also a major issue to be confronted. In order to address these issues, a proper understanding of the supercapacitor concept and device architecture is mandatory. This section discusses the major concepts, device assembly designs, and the significant modifications to be adopted for the fabrication of high-performance supercapacitors.

As mentioned in previous chapters, the capacitor assembly consists of two electrodes, a cathode and an anode, a separator, and an electrolyte. The properties of the material and electrolyte employed, and the mechanism involved decides the performance of supercapacitors. In EDLCs, the charge storage performance relies on the electrostatic interactions of ions at the electrode/electrolyte interface and the separation of charge takes place in the Helmholtz double layer. In EDLCs, carbonaceous electrode materials are mostly employed. In PC designs, the fast faradaic redox reactions or intercalation processes take place at the electrode or in the vicinity of the electrode and results in charge accumulation. In this case, doped and undoped metal oxides, conducting polymers, nanomaterials, and other advanced state-of-the-art materials are used for the fabrication. The third type of supercapacitor is the hybrid supercapacitor constituted of the integration of the previous two which exhibits both faradaic and non-faradaic types of modus operandi. Such hybrid supercapacitors offer high energy density, a wide potential window, and high upper-limit voltage. Several permutations and combinations of electrode configurations can be achieved with the hybrid supercapacitors either as symmetric or asymmetric assembly. In general, the supercapacitor performance depends on many factors such as double-layer thickness characteristics of active electrode materials including morphology, size/amount, microstructure, surface area, porosity, synthesis method adopted, and the electrolyte employed. In the design of highly efficient supercapacitors, we must consider the requisites and type of electrode/electrolyte materials needed for a particular application.

4.4.1 METHODS FOR TUNING SUPERCAPACITOR PERFORMANCE

4.4.1.1 Nanostructuring of Electrodes

Normally, the active material used for the device architecture should possess a high surface area and sufficient and well-defined pore geometry. The specific capacitance and ionic conductivity mainly hinge on these parameters. A high surface area provides sufficient space for the electrochemical reactions and charge storage, thereby increasing the specific capacitance. The higher pore size distribution of the current collector provides efficient superhighways for charge transport and leads to increased ionic conductivity. The precise control of pore size and provident design strategies clearly indicates that more open structures assist higher ion transport. A number of technologies have been developed for tuning the morphological specifications. Among them, nano structuring is the core technique used for the surface modifications. The progressing nanotechnology is a blessing for the device fabrication. A variety of nanostructures such as nanotubes, nanowires, nanospheres, and nanorods were implemented for supercapacitor applications. Selection of the appropriate structure for particular application is vital. Apart from the higher surface area and well-defined geometry, the excellent stability of nanomaterials offers increased cyclic stability and battery life. Highly stable nanostructures also act as a substrate for many electrode designs and the highly porous nanostructured substrate is beneficial for depositing less stable active materials in its well-defined pores, thereby increasing the specific capacitance value.

4.4.1.2 Chemical Activation of Active Electrode Material

In the chemical activation method, the precursor solutions are impregnated with a chemical reagent followed by a heating process under an inert atmosphere. As we know, a vast range of materials is used as electrode materials, including several carbonaceous materials, metal oxides, conducting polymers, etc. Among them, activated carbons are the most used electrode materials which possess an EDLC and a

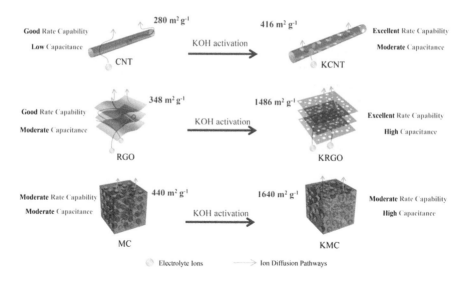

FIGURE 4.2 Schematic illustration of the ion diffusion pathways and specific surface areas of carbon nanotube, reduced graphene oxide, and mesoporous carbon with KOH activation. The performance of capacitance and the capability are also evaluated according to the result of electrochemical tests. Reproduced from Xiong et al. [55].

faradaic mechanism. As per the environmentally friendly concept, biomass-derived activated carbon production is gaining wide consideration. A number of biopolymer-derived methods i being developed today where chemical activation is done with the use of many reagents including $ZnCl_2$, H_3PO_4, KOH, NaOH, etc., to enhance morphology, pore distribution, and thus capacitance [53, 54]. Xiong et al. investigated chemical activation of various activated carbons and studied the effect of spatial characteristics of carbon precursors such as 1D carbon nanotube, 2D reduced GO, and 3D mesoporous carbon on the chemical activation with KOH. The results provide significant information on chemical activation of carbon materials for optimizing the supercapacitive performance and is illustrated in Figure 4.2 [55]. Numerous studies are in progress to develop highly efficient carbon materials with different chemical activation methods as a reagent [56–58]. Elemental doping such as nitrogen doping, boron doping, phosphorous doping, sulfur doping are also efficient strategies to obtain high performance capacitor electrodes with improved morphology, porosity, and chemical stability. The exfoliation method has also proven to be an effective strategy for enhancing capacitor performance. In the exfoliation method, rich active sites were introduced on a layered structure such as graphene for charged particle diffusion, thereby increasing the capacitance [59].

Transition metal oxides are considered as a major supercapacitor electrode material, with their remarkable theoretical specific capacitance for faradaic charge storage compared to carbon electrodes. However poor electrical conductivity of most of the metal oxides, thereby reducing rate capability and capacitance, is a huge challenge [60, 61]. A variety of chemical methods has been adopted to activate electrodes. The introduction of oxygen vacancies and elemental doping acquired spectacular recognition for improving poor electrical conductivity performance. Chemical exfoliation methods are also a chief strategy to improve supercapacitor performance. Controlled formation of oxygen vacancies in metal oxide materials has been established as an effective strategy to increase the carrier densities and consequently electrical conductivities. The chemical reduction and polarization method, together with metal and nonmetal doping, result in oxygen vacancy creation. The introduction of defects and element doping can greatly boost the electrical conductivity of transition metal oxides, however, the effects on the ion diffusion rate in bulk electrodes is very weak. Integration with unique 2D metal oxide structures via chemical exfoliation demonstrates prodigious capacity for improving the ion diffusion problem [59]. Chemical activation is found to be more efficient and convenient in synthesizing high quality electrodes with superior electrochemical performance.

4.4.1.3   Physical Activation of Active Electrode Material

In a physical activation strategy, the precursor is carbonized at elevated temperatures by passing oxidizing gases like CO_2, H_2O vapors, etc. [62]. Physical activation with CO_2 adsorption can be controlled without any difficulty and can produce microporous carbon. Activation with H_2O brings out a mesoporous structure with evenly distributed pores. The direct and indirect CO_2 activation results in a dominant micro/nanoporous structure with a limited number of large pores which is extremely favorable for supercapacitor applications in terms of ion storage and transport. In a typical procedure, the CO_2 gas activation produces the best physical properties for the carbon electrode with a smaller density and higher carbon content than with H_2O [63]. The CO_2-derived porous carbons with a huge volume of narrow small pores enhances both the energy and power density of supercapacitor devices. The presence of surface functional groups also plays a vital role in improving the electrochemical performance.

As we discussed earlier, oxygen vacancies can also be created by thermal annealing of metal oxides in an oxygen-deficient environment, such as nitrogen or argon. For instance, Salari et al. developed thermally treated rutile TiO_2 from anatase TiO_2 in an inert atmosphere at an elevated temperature of 600°C. The lower partial pressure of oxygen gas in the environment leads to oxygen-deficient vacancies in transition metal oxides [64]. The areal capacitance of TiO_2 treated in argon is found to be 521 μF cm^{-2} and was higher than the air annealed (30 μF cm^{-2}) at the scan rate of 1 mV s^{-1}. Hydrogen treatment can also substantially increase the donor density, and hence the specific capacitance of electrode material.

4.5   CONCLUSION

This chapter discusses the electrode processes, such as double-layer capacitance and pseudocapacitance. Numerous materials have been employed for supercapacitive electrode fabrication, such as carbonaceous materials, transition metal oxides, etc. Each material has a different effect on supercapacitor performance. The selection of a particular material defines the device mechanism. In addition, structure and porosity also have a very crucial role in deciding supercapacitor performance. A high surface area and well-defined porous geometry enhance capacitor performance and results in high conductivity and specific capacitance with improved stability. The chapter also discusses the strategies adopted for improving electrochemical performance. As technology is progressing, the research on supercapacitors gains momentum day by day.

REFERENCES

1. Attia, S. Y., Mohamed, S. G., Barakat, Y. F., Hassan, H. H., & Zoubi, W. A. (2022). Supercapacitor electrode materials: Addressing challenges in mechanism and charge storage. *Reviews in Inorganic Chemistry*, *42*(1), 53–88.
2. Ghosh, A., & Lee, Y. H. (2012). Carbon-based electrochemical capacitors. *ChemSusChem*, *5*(3), 480–499.
3. Liu, Y., Jiang, S. P., & Shao, Z. (2020). Intercalation pseudocapacitance in electrochemical energy storage: Recent advances in fundamental understanding and materials development. *Materials Today Advances*, *7*, 100072.
4. Electrochemical Supercapacitors_ Scientific Fundamentals and Technological Applications (CPENTalk.com).pdf. (n.d.).
5. Herrero, E., Buller, L. J., & Abruña, H. D. (2001). Underpotential deposition at single crystal surfaces of Au, Pt, Ag and other materials. *Chemical Reviews*, *101*(7), 1897–1930.
6. Augustyn, V., Simon, P., & Dunn, B. (2014). Pseudocapacitive oxide materials for high-rate electrochemical energy storage. *Energy & Environmental Science*, *7*(5), 1597.
7. Yoo, H. D., Li, Y., Liang, Y., Lan, Y., Wang, F., & Yao, Y. (2016). Intercalation pseudocapacitance of exfoliated molybdenum disulfide for ultrafast energy storage. *ChemNanoMat*, *2*(7), 688–691.
8. Chodankar, N. R., Pham, H. D., Nanjundan, A. K., Fernando, J. F. S., Jayaramulu, K., Golberg, D., … Dubal, D. P. (2020). True meaning of pseudocapacitors and their performance metrics: Asymmetric versus hybrid supercapacitors. *Small*, *16*(37), 2002806.
9. Lian, Q., Zhou, G., Liu, J., Wu, C., Wei, W., Chen, L., & Li, C. (2017). Extrinsic pseudocapacitve Li-ion storage of SnS anode via lithiation-induced structural optimization on cycling. *Journal of Power Sources*, *366*, 1–8.
10. Lima-Tenório, M. K., Ferreira, C. S., Rebelo, Q. H. F., Souza, R. F. B.de, Passos, R. R., Pineda, E. A. G., & Pocrifka, L. A. (2018). Pseudocapacitance properties of Co_3O_4 nanoparticles synthesized using a modified sol-gel method. *Materials Research*, *21*(2), 1–7.
11. Simon, P., Gogotsi, Y., & Dunn, B. (2014). Where do batteries end and supercapacitors begin? *Science*, *343*(6176), 1210–1211.
12. Trasatti, S. (1983). Physical, chemical and structural aspects of the electrode/solution interface. *Electrochimica Acta*, *28*(8), 1083–1093.
13. Zheng, J. P., Cygan, P. J., & Jow, T. R. (1995). Hydrous ruthenium oxide as an electrode material for electrochemical capacitors. *Journal of The Electrochemical Society*, *142*(8), 2699–2703.
14. [. Yuan, C., Hou, L., Li, D., Yang, L., & Li, J. (2012). Enhanced supercapacitance of hydrous ruthenium oxide/mesocarbon microbeads composites toward electrochemical capacitors. *International Journal of Electrochemistry*, *2012*, 1–7.
15. Zhang, K., Han, X., Hu, Z., Zhang, X., Tao, Z., & Chen, J. (2015). Nanostructured Mn-based oxides for electrochemical energy storage and conversion. *Chemical Society Reviews*, *44*(3), 699–728.

16. Ghodbane, O., Pascal, J.-L., & Favier, F. (2009). Microstructural effects on charge-storage properties in MnO_2 -based electrochemical supercapacitors. *ACS Applied Materials & Interfaces, 1*(5), 1130–1139.
17. Toupin, M., Brousse, T., & Bélanger, D. (2004). Charge storage mechanism of MnO_2 electrode used in aqueous electrochemical capacitor. *Chemistry of Materials, 16*(16), 3184–3190.
18. Yan, W., Kim, J. Y., Xing, W., Donavan, K. C., Ayvazian, T., & Penner, R. M. (2012). Lithographically patterned gold/manganese dioxide core/shell nanowires for high capacity, high rate, and high cyclability hybrid electrical energy storage. *Chemistry of Materials, 24*(12), 2382–2390.
19. Brousse, T., Toupin, M., Dugas, R., Athouël, L., Crosnier, O., & Bélanger, D. (2006). Crystalline MnO_2 as possible alternatives to amorphous compounds in electrochemical supercapacitors. *Journal of The Electrochemical Society, 153*(12), A2171.
20. Chernova, N. A., Roppolo, M., Dillon, A. C., & Whittingham, M. S. (2009). Layered vanadium and molybdenum oxides: Batteries and electrochromics. *Journal of Materials Chemistry, 19*(17), 2526.
21. Amatucci, G. G., Badway, F., Singhal, A., Beaudoin, B., Skandan, G., Bowmer, T., … Jaworski, R. (2001). Investigation of yttrium and polyvalent ion intercalation into nano-crystalline vanadium oxide. *Journal of The Electrochemical Society, 148*(8), A940.
22. Dong, W. (1999). Electrochemical properties of high surface area vanadium oxide aero-gels. *Electrochemical and Solid-State Letters, 3*(10), 457.
23. Kim, H.-S., Cook, J. B., Tolbert, S. H., & Dunn, B. (2015). The development of pseu-docapacitive properties in nanosized-MoO_2. *Journal of the Electrochemical Society, 162*(5), A5083–A5090.
24. Wang, Y., Song, Y., & Xia, Y. (2016). Electrochemical capacitors: Mechanism, mate-rials, systems, characterization and applications. *Chemical Society Reviews, 45*(21), 5925–5950.
25. Augustyn, V., Come, J., Lowe, M. A., Kim, J. W., Taberna, P.-L., Tolbert, S. H., … Dunn, B. (2013). High-rate electrochemical energy storage through Li^+ intercalation pseudocapacitance. *Nature Materials, 12*(6), 518–522.
26. Okubo, M., Hosono, E., Kim, J., Enomoto, M., Kojima, N., Kudo, T., … Honma, I. (2007). Nanosize effect on high-rate Li-ion intercalation in $LiCoO_2$ electrode. *Journal of the American Chemical Society, 129*(23), 7444–7452.
27. Zhao, Y., Han, C., Yang, J., Su, J., Xu, X., Li, S., … Zhang, Q. (2015). Stable alkali metal ion intercalation compounds as optimized metal oxide nanowire cathodes for lithium batteries. *Nano Letters, 15*(3), 2180–2185.
28. Wang, J., Polleux, J., Brezesinski, T., Tolbert, S., & Dunn, B. (2008). The Pseudocapacitance Behaviors of TiO2 (Anatase) Nanoparticles. *ECS Transactions, 11*(31), 101–111.
29. Kudo, T., Obayashi, H., & Gejo, T. (1975). Electrochemical Behavior of the Perovskite-Type $Nd^{1-x} Sr^x CoO_3$ in an Aqueous Alkaline Solution. *Journal of The Electrochemical Society, 122*(2), 159–163.
30. Mefford, J. T., Hardin, W. G., Dai, S., Johnston, K. P., & Stevenson, K. J. (2014). Anion charge storage through oxygen intercalation in LaMnO3 perovskite pseudocapacitor electrodes. *Nature Materials, 13*(7), 726–732.
31. Chiabrera, F., Baiutti, F., Diercks, D., Cavallaro, A., Aguadero, A., Morata, A., & Tarancón, A. (2022). Visualizing local fast ionic conduction pathways in nanocrystal-line lanthanum manganite by isotope exchange-atom probe tomography. *Journal of Materials Chemistry A*, 10.1039.D1TA10538H.
32. Bleda-Martínez, M. J., Maciá-Agulló, J. A., Lozano-Castelló, D., Morallón, E., Cazorla-Amorós, D., & Linares-Solano, A. (2005). Role of surface chemistry on electric double layer capacitance of carbon materials. *Carbon, 43*(13), 2677–2684.

33. Ma, F., Ding, S., Ren, H., & Liu, Y. (2019). Sakura-based activated carbon preparation and its performance in supercapacitor applications. *RSC Advances*, *9*(5), 2474–2483.
34. González, A., Goikolea, E., Barrena, J. A., & Mysyk, R. (2016). Review on supercapacitors: Technologies and materials. *Renewable and Sustainable Energy Reviews*, *58*, 1189–1206.
35. Zhi, M., Xiang, C., Li, J., Li, M., & Wu, N. (2013). Nanostructured carbon–metal oxide composite electrodes for supercapacitors: A review. *Nanoscale*, *5*(1), 72–88.
36. Qu, D. (2014). Fundamental principals of battery design: Porous electrodes (pp. 14–25). Presented at the Review On Electrochemical Storage Materials And Technology: Proceedings of the 1st International Freiberg Conference on Electrochemical Storage Materials, Freiberg, Germany.
37. Zhang, L. L., Gu, Y., & Zhao, X. S. (2013). Advanced porous carbon electrodes for electrochemical capacitors. *Journal of Materials Chemistry A*, *1*(33), 9395.
38. Choi, B. G., Yang, M., Hong, W. H., Choi, J. W., & Huh, Y. S. (2012). 3D Macroporous graphene frameworks for supercapacitors with high energy and power densities. *ACS Nano*, *6*(5), 4020–4028.
39. Candelaria, S. L., Garcia, B. B., Liu, D., & Cao, G. (2012). Nitrogen modification of highly porous carbon for improved supercapacitor performance. *Journal of Materials Chemistry*, *22*(19), 9884.
40. Zhou, J., Zhang, Z., Xing, W., Yu, J., Han, G., Si, W., & Zhuo, S. (2015). Nitrogen-doped hierarchical porous carbon materials prepared from meta-aminophenol formaldehyde resin for supercapacitor with high rate performance. *Electrochimica Acta*, *153*, 68–75.
41. Jung, S. M., Mafra, D. L., Lin, C.-T., Jung, H. Y., & Kong, J. (2015). Controlled porous structures of graphene aerogels and their effect on supercapacitor performance. *Nanoscale*, *7*(10), 4386–4393.
42. Yu, Z., Tetard, L., Zhai, L., & Thomas, J. (2015). Supercapacitor electrode materials: Nanostructures from 0 to 3 dimensions. *Energy & Environmental Science*, *8*(3), 702–730.
43. Lang, X., Hirata, A., Fujita, T., & Chen, M. (2011). Nanoporous metal/oxide hybrid electrodes for electrochemical supercapacitors. *Nature Nanotechnology*, *6*(4), 232–236.
44. Lai, X., Halpert, J. E., & Wang, D. (2012). Recent advances in micro-/nano-structured hollow spheres for energy applications: From simple to complex systems. *Energy Environ. Sci.*, *5*(2), 5604–5618.
45. You, B., Yang, J., Sun, Y., & Su, Q. (2011). Easy synthesis of hollow core, bimodal mesoporous shell carbon nanospheres and their application in supercapacitor. *Chemical Communications*, *47*(45), 12364.
46. Wang, K., Huang, J., & Wei, Z. (2010). Conducting polyaniline nanowire arrays for high performance supercapacitors. *The Journal of Physical Chemistry C*, *114*(17), 8062–8067.
47. Kim, B., Chung, H., & Kim, W. (2010). Supergrowth of aligned carbon nanotubes directly on carbon papers and their properties as supercapacitors. *The Journal of Physical Chemistry C*, *114*(35), 15223–15227.
48. Stoller, M. D., Park, S., Zhu, Y., An, J., & Ruoff, R. S. (2008). Graphene-based ultracapacitors. *Nano Letters*, *8*(10), 3498–3502.
49. Huang, J., Sumpter, B. G., & Meunier, V. (2008). Theoretical model for nanoporous carbon supercapacitors. *Angewandte Chemie International Edition*, *47*(3), 520–524.
50. Huang, J., Sumpter, B. G., & Meunier, V. (2008). A universal model for nanoporous carbon supercapacitors applicable to diverse pore regimes, carbon materials, and electrolytes. *Chemistry: A European Journal*, *14*(22), 6614–6626.

51. Deng, M.-J., Ho, P.-J., Song, C.-Z., Chen, S.-A., Lee, J.-F., Chen, J.-M., & Lu, K.-T. (2013). Fabrication of Mn/Mn oxide core–shell electrodes with three-dimensionally ordered macroporous structures for high-capacitance supercapacitors. *Energy & Environmental Science*, *6*(7), 2178.

52. https://www.cap-xx.com/resource/the-supercapacitor-advantage/. (n.d.).

53. Boujibar, O., Ghamouss, F., Ghosh, A., Achak, O., & Chafik, T. (2019). Activated carbon with exceptionally high surface area and tailored nanoporosity obtained from natural anthracite and its use in supercapacitors. *Journal of Power Sources*, *436*, 226882.

54. Molina-Sabio, M., & Rodríguez-Reinoso, F. (2004). Role of chemical activation in the development of carbon porosity. *Colloids and Surfaces A: Physicochemical and Engineering Aspects*, *241*(1–3), 15–25.

55. Xiong, R., Zhang, Y., Zhou, W., Xia, K., Sun, Q., Chen, G., … Zhou, C. (2020). Chemical activation of carbon materials for supercapacitors: Elucidating the effect of spatial characteristics of the precursors. *Colloids and Surfaces A: Physicochemical and Engineering Aspects*, *597*, 124762.

56. Ciftyurek, E., Bragg, D., Oginni, O., Levelle, R., Singh, K., Sivanandan, L., & Sabolsky, E. M. (2019). Performance of activated carbons synthesized from fruit dehydration biowastes for supercapacitor applications. *Environmental Progress & Sustainable Energy*, *38*(3).

57. Taer, E., Febriyanti, F., Mustika, W. S., Taslim, R., Agustino, A., & Apriwandi, A. (2021). Enhancing the performance of supercapacitor electrode from chemical activation of carbon nanofibers derived Areca catechu husk via one-stage integrated pyrolysis. *Carbon Letters*, *31*(4), 601–612.

58. Taer, E., Sugianti, T. E., Apriwandi, Rini, A. S., Malik, U., & Taslim, R. (2021). Low-cost activated carbon bio-wasted-based for enhanced capacitive properties of symmetric supercapacitor. *Journal of Physics: Conference Series*, *2049*(1), 012007.

59. Kou, T., Yao, B., Liu, T., & Li, Y. (2017). Recent advances in chemical methods for activating carbon and metal oxide based electrodes for supercapacitors. *Journal of Materials Chemistry A*, *5*(33), 17151–17173.

60. Dong, C., Wang, Y., Xu, J., Cheng, G., Yang, W., Kou, T., … Ding, Y. (2014). 3D binder-free $Cu_2O@Cu$ nanoneedle arrays for high-performance asymmetric supercapacitors. *Journal of Materials Chemistry A*, *2*(43), 18229–18235.

61. Yu, M., Wang, Z., Han, Y., Tong, Y., Lu, X., & Yang, S. (2016). Recent progress in the development of anodes for asymmetric supercapacitors. *Journal of Materials Chemistry A*, *4*(13), 4634–4658.

62. Shrestha, D., & Rajbhandari, A. (2021). The effects of different activating agents on the physical and electrochemical properties of activated carbon electrodes fabricated from wood-dust of Shorea robusta. *Heliyon*, *7*(9), e07917.

63. Taer, E., Apriwandi, Yusriwandi, Mustika, W. S., Zulkifli, Taslim, R., … Dewi, P. (2018). Comparative study of CO2 and H2O activation in the synthesis of carbon electrode for supercapacitors (p. 030036). Presented at the 1st International Conference and Exhibition on Powder Technology Indonesia (ICePTi) 2017, Jatinangor, Indonesia.

64. Salari, M., Konstantinov, K., & Liu, H. K. (2011). Enhancement of the capacitance in TiO2 nanotubes through controlled introduction of oxygen vacancies. *Journal of Materials Chemistry*, *21*(13), 5128.

5 Design Considerations

Mijun Chandran, Asha Raveendran,
Mari Vinoba, and Margandan Bhagiyalakshmi

CONTENTS

5.1 INTRODUCTION

The depletion of fossil fuels accompanied by their harmful impact on the environment has caused a shift of attention to the development of sustainable energy storage and conversion devices worldwide [1–4]. The different energy storage devices include supercapacitors and batteries, and a Ragone plot illustrates their energy and power densities; they are considered the leading electrochemical energy storage devices. In fact, batteries have a high energy density and are hence recommended for applications in many electronic devices. However, the relatively slow electron and ion transport and resistive/heat losses of batteries makes them less competitive in many applications [5–7]. On the other hand, supercapacitors have a fast-charging ability,

DOI: 10.1201/9781003258384-5

stability of more than 10,000 cycles, and high power density compared to batteries, which have triggered the growth in interest. The commercialization of supercapacitors has been limited due to poor energy density (~5 Wh kg^{-1}), which is comparatively lower than batteries (up to 200 Wh kg^{-1}) [8]. Extensive research is underway to produce a supercapacitor with high energy density, power density, and cyclic stability. The crucial factors that influence the efficiency of supercapacitors are the design, operation, and manufacturing of these energy storage devices [9–12]. Systematic design and integration of electrodes, electrolytes, and current collectors could alter the performance of electrochemical supercapacitors and batteries to a greater extent. Thus, supercapacitor manufacturers can apply battery engineering to make the process better and cost effective [13]. The current supercapacitors are expensive for adequate applications; if supercapacitors are developed with inexpensive materials that exhibit excellent specific capacitance, it could make the manufacturing process more viable, driven by the need to be innovative, decrease manufacturing costs that limit competition, and improve sales [14–15]. In order to account for the above-mentioned, appropriate design, electrode, and electrolyte material are other factors required for the construction of a supercapacitor with high efficiency are discussed in this chapter.

5.2 SUPERCAPACITOR SYSTEM DESIGN CONSIDERATIONS

Factors such as cell polarity, cyclic stability, and heat management give information about the efficiency and performance of the device. However, more complex solutions are to be looked into to realize their optimal performance. Among the many design considerations, the important ones include cell voltage, frequency response, humidity, lifetime and cycle charging, ambient temperature effects, efficiency, and interconnection.

5.2.1 CELL VOLTAGE

Compared to batteries, supercapacitors work over a wider potential window, from rated voltage to zero volts. The voltage rating depends on the series and size, which is acquired by the electrochemical stability of the electrode and electrolyte material. The amount of energy stored is given by the equation $E = \frac{1}{2} Cv^2$, where energy density is proportional to the square of the cell potential dependent on the electrolyte potential window [16–18]. Thus, electrolytes have an essential effect on the cell voltage window; while aqueous electrolytes have a possible window around 1–1.5 V, organic electrolytes attain a high operational voltage of around 3 V. A typical electrochemical supercapacitor has several 2.7 V single cells arranged in series for a continuous power supply. An extensive functional potential window increases the overall efficiency of both storage and supply of energy. However, if the cell is operated above the voltage for a long time, cell life is affected and reduced, leading to electrolyte decomposition [19]. The damage done depends on the voltage applied and the period the supercapacitor is exposed to the high voltage scenario. The occasional spike, however, does not immediately affect the device. The supercapacitor can acquire the maximum energy if it's discharged to zero, i.e., the system possesses a minimum threshold voltage that ultimately limits the lower voltage at which the capacitor can discharge [20–22]. Currently, stored energy up to 75% can be utilized when the supercapacitor is discharged from the initial charge potential V initial to one-half of V initial [23].

5.2.2 FREQUENCY RESPONSE

The time constant (τ) denotes the time to acquire 63.2% of energy from a non-charged state or discharge 32.8% from an ultimately charged state. The factor frequency response for an electrochemical supercapacitor is entirely dependent on its calculated time constant [24, 25]. Based on the electrode, electrolyte material, and cell configuration, the frequency response is in milliseconds to seconds. They ultimately contribute to the capacitance and the equivalent series resistance (ESR) [24]. Thus, increasing the capacitances and decreasing the resistance make supercapacitors applicable in pulse power applications requiring fast responses with limited power losses.

5.2.3 AMBIENT TEMPERATURE

One of the advantages of supercapacitors is their vast temperature capability, i.e., their operating temperature ranges from −40 to around 65°C, making them suitable for different regional and environmental conditions [26]. Organic electrolytes have a lower freezing point and are considered more favourable compared to aqueous electrolytes. The changing temperature does not affect the capacitance as energy storage or charge is not a chemical reaction. However, the resistance is co-dependent on the mobility of ions within the electrolyte. The mobility decreases as the temperature drops, thereby resulting in higher resistance [27]. The high operating temperature may affect the capacitance, diminishing the efficiency and altering the positive effect of ESR reduction. The surface area to volume ratio is also an essential factor. It is observed that the electrochemical supercapacitor stack is affected by the temperature effects due to their higher surface area-to-volume ratio [28, 29].

5.2.4 POLARITY

Electrochemical supercapacitors are differentiated as electric double-layer supercapacitors, pseudo supercapacitors, and hybrid supercapacitors. Electric double-layer supercapacitors are symmetric devices where both the anode and cathode are composed of the same material and hence theoretically have no polarity [30]. For consistency purposes, while manufacturing, the intended terminal polarity is indicated by using stainless steel casing. If the supercapacitor is conditioned for charge in a particular direction, switching its path can lower the life; however, it would not cause a severe failure. In the case of the asymmetric supercapacitors, the anode and cathode materials are different, and thus, polarity matters. Other electrodes have different chemical reactions leading to varying potentials and different polarities. The marking of contradiction is necessary for asymmetric supercapacitors to indicate their performance in specific operation voltage windows.

5.2.5 LIFETIME AND CYCLE CHARGING

Compared to batteries, a supercapacitor has unlimited life, ultimately affecting the cell potential and temperature. The repeated charging and discharge at constant current may lead to the exponential decay of capacitance, causing a rise in internal resistance [31]. Based on a particular datasheet, the lifecycle is deemed to end if the capacitance drops by 30% and the internal resistance doubles. Their performance falls below the expected standard of the application requirements that could vary for different datasheets.

In the case of cycle charging, commercial supercapacitors have a life more significant than 500,000 cycles with slight performance loss [32]. The charge to discharge current ratio is called coulombic efficiency, which may reach 100% in extreme winds. For practical applications, the charge and discharge currents are maintained to be the same to have maximum coulombic efficiency. This efficiency, however, does not entirely correspond to maximum energy efficiency.

5.2.6 HUMIDITY

Supercapacitors can work in high levels of humidity. However, the mainstream products are advised to be used within six months after the opening of the vacuum seal, as their exposure to moisture can lead to oxidation of the soldered leads, leading to corrosion. If all the collected cells are not used, the oxidized ends can be removed carefully using an abrasive emery cloth with proper precautions [33].

5.2.7 EFFICIENCY

During charging and discharge, supercapacitors have almost the same efficiency as batteries. The efficiency losses are only contributed to by internal resistance (IR) causing IR drop during cycling. Though in most cases, the efficiency of supercapacitors is around 90%, only at high current and power applications is the efficiency affected.

5.3 SINGLE CELL MANUFACTURING

5.3.1 ELECTRODE

The source for energy storage and one of the significant parts of supercapacitor devices is the electrode. The interaction between the electrode/electrolyte interface is responsible for the charge storage. The area, porosity, and conductivity of the electrode affect the energy density drastically. As the active surface area increases, the energy density increases as the number of ions stored during the formation of the Helmholtz double-layer also increases. The most popular candidate for the electrode is activated carbon, as they show sufficient capacitance in both organic and aqueous electrolytes [34–36].

The electrode is a significant component of the supercapacitor system and is a site for energy storage. At the electrode/electrolyte interface, the ions will interact with the electrode material and are stored at the active sites of the electrode. The principal parameters of a supercapacitor electrode are porosity, area, and conductivity. Such electrode parameters have a considerable influence over the energy density [37]. Optimization of such parameters will lead to high-functioning supercapacitors and the application in existing electronic devices such as laptops and mobile phones. For the supercapacitor electrode, the amount of energy stored depends mainly on the electrode surface area; considerable research is going on globally in search of materials with a large surface area for their application in energy storage devices. With the increase in the active surface area, the number of ions that can be stored by forming a Helmholtz double layer will also increase, which leads to higher energy density. Activated carbon is a good candidate for electrode material as it shows adequate capacitance with both aqueous and organic electrolytes. Significant properties of carbon-based electrodes like high conductivity, high surface area ($1–3000$ m^2/g), good corrosion resistance, high-temperature stability, controlled pore structure, compatibility for composite formation, and relatively low cost make them eligible support for energy storage application [38–40].

The increase in surface area-specific capacitance also increases proportionally. However, a large surface area does not always ensure high specific capacitance. It depends largely on the accessible surface area and pore size of the electrode material. Electric double-layer capacitors (EDLCs) are usually made of activated carbon, graphene, and carbon nanotubes (CNTs) [39]. In comparison, pseudocapacitors are generally composed of metal oxides and conducting polymers. High specific area electrode materials are commonly carbon-based materials like carbon fibres, aerogels, and CNTs. They have high conductivity, high surface area, high temperature stability, high corrosion resistance, and good compatibility. These materials have a high specific place with meso- or macropores, which facilitate rapid ion transport throughout the carbonaceous matrix, making them have excellent electrochemical performance.

Activated carbon is prepared via the thermal or chemical treatment of carbonaceous material to improve its surface area. The porous structure of activated carbon consists of micropores (diameter <2 nm), mesopores (diameter $2–50$ nm), and macropores (diameter >50 nm). Smaller micropores cannot support the diffusion of an electrolyte, thus inhibiting certain pores from charge storage. Activated carbon, being cheap compared to other carbon-based materials, is used extensively in EDLCs.

Carbon black particles are spherical, and the size is in the range of colloidal particles. They are frequently used as filling materials in the electrodes of both batteries and supercapacitors. They have high porosity and chemically clean surface area, which enhance the electrochemical activity of EDLCs. The specific capacitance of carbon black is about 250 Fg^{-1} for EDLCs.

Also, carbon fibres synthesized from cellulose, polyacrylonitrile, pitch-based materials, and phenolic resin, thermosetting organic materials with high adsorption capacities and high adsorption rates are also recognized as promising electrode materials for ELDCs. Due to the presence of pores, good accessibility to active sites is achieved. However, the drawback is that the contact resistance between individual fibres is high, and their cost is much higher than other carbonaceous products.

Carbon aerogels are synthesized by the polycondensation of resorcinol and formaldehyde and followed by pyrolysis. They have a large surface area, uniform size, higher density, and excellent conductivity. Carbon aerogel activation can improve the active surface area via thermal, electrochemical, and chemical vapour impregnation.

CNTs are synthesized via the catalytic decomposition of hydrocarbons. They are of two types: single-walled CNTs and multi-walled CNTs. The surface area of CNTs compared to activated carbon is much lower; however, they have a specific capacitance of 15–18 Fg^{-1} which can be enhanced up to 130 Fg^{-1} when the surface texture is altered via oxidative treatment. Their conductivity, surface area, thermal, and chemical stability make them suitable as electrode materials for supercapacitors.

Transition metal oxides and conducting polymers can have fast reversible redox reactions on their surface, making them suitable electrode materials for pseudocapacitors. Moreover, compared to carbon-based materials, these materials exhibit higher specific capacitance. Due to their high capacitance at small resistance, metal oxides like RuO_2, NiO, Co_3O_4, and MnO_2 are studied as electrode materials. The specific capacitance of hydrous RuO2 in sulphuric acid was 750 Fg^{-1} owing to the fast transfer of electrons and protons absorbed on the electrochemical surface. The specific capacitance of about 256 Fg^{-1} is observed for NiO, prepared by the sol-gel technique. Compared to other metal oxides, MnO_2 is environmentally friendly and less expensive [36, 37].

In addition, various conducting polymers, polyaniline (PANI), polyethyleneimine (PEI), polypyrrole (PPY), Poly (3,4-ethylene dioxythiophene) (PEDOT), polythiophene (PTH), and poly (p-phenylene vinylene) (PPV), are also used as electrode materials in supercapacitor applications [38]. They enhance electrical conductivity due to the presence of overlapping π-conjugated polymer chains in the backbone. They have higher conductivity, high capacitance, and low equivalent series resistance than carbon-based material, and they are more environmentally friendly and cost effective than transition metal oxides. The doping of ions onto polymers has been shown to enhance the capacitance of the conductivity polymers significantly. Factors such as type of dopant, monomers, electrolyte, pH, and deposition condition play an important role in influencing the capacitance value of the conducting polymer film. Due to lower cyclic stability as sound degradation caused by overoxidation, conducting polymer as an electrode material is limited.

5.3.2 Electrolyte

An electrolyte is usually composed of liquid gel or solid, which facilitates the movement of ions/electrons to and from the electrode under the applied voltage. The electrolyte influences the performance of the supercapacitor in terms of energy/power density. The parameters that influence the electrolyte performance are ion size, ionic conductivity, chemical reactivity, solvent molecule, and viscosity [40–43]. The electrolyte should be stable both chemically and thermally. In general, inert electrolytes are used in the EDLC system to increase the long cycle stability of the supercapacitor system. For a fast reversible redox reaction in pseudocapacitance-based supercapacitors, the electrolyte must have chemical compatibility with electrode material and electrode functional groups solvent molecules [41]. Ionic conductivity relies on the concentration of electrolytes; conductivity decreases as the number of ions decreases at a low concentration of electrolytes. At higher electrolyte concentrations, collisions occur among the solvated ions due to their dense distribution, which influences the ionic mobility, and thus the power density is also affected [42]. Therefore, ionic conductivity depends on the type of solvent and their size as well as their concentration.

The different types of electrolytes are aqueous, organic, and ionic electrolytes. Solid electrolytes are solids or gels with ions and are easier to synthesize but have much lower conductivity than the aqueous electrolyte [41–47]. The aqueous electrolyte has only operational cell voltage up to 1 V, which decreases the energy density. Beyond 1.2 V, supercapacitor performance declined due to the dissociation of water into hydrogen and oxygen, increasing electrolyte viscosity, and decreased ion mobility. They have low specific resistance and high power density. The commonly used aqueous electrolytes are acidic and essential. A high concentration of electrolytes improves the conductivity and reduces the equivalent series resistance. Concentrated sulphuric acid is a prevalent electrolyte for RuO_2 to avoid electrolyte depletion problems during charging. Mild aqueous KCl electrolyte is also used for MnO_2 electrodes. Other commonly used electrolytes are potassium electrolytes, NaOH and NaCl. They are less expensive, and their purification and drying processes are comparatively less stringent and cheaper [41, 42].

On the other hand, organic electrolytes are composed of acetonitrile or propylene carbonate, which operates in the voltage range of 2.5–3 V. However, the specific resistance of organic electrolytes is much higher than aqueous electrolytes and reduces the low power density. They require stringent purification and drying production, which may otherwise lead to self-discharge due to recombination reactions. They are comparatively more expensive and have higher resistivity than the aqueous electrolytes. The electrode should have a greater pore size as the organic molecules would have a considerable extent; this ultimately decreases the power density of the supercapacitors. Among the available electrolytes, tetraethyl ammonium salts, acetonitrile, and N-methyl-2-pyrrolidone (NMP) are more suitable candidates for electrochemical supercapacitors due to their excellent conductivity, solubility, and higher dissociation potential. Ionic electrolytes are bulky organic anions and cations, and they don't have solvent molecules, unlike others that dissolve salts in the molecular solvent. Supercapacitors employed with ionic electrolytes can exhibit a large potential window up to 3.4 V and hence show high energy and power density. Ionic electrolytes are thermally/chemically stable, with negligible vapour pressure, and are generally liquid. But being highly expensive, the use of ionic liquid-based supercapacitors is limited for commercialization.

5.3.3 Separator

Separators are ion-permeable that allow ionic charge transfer to occur but hinder the electrical contact between two electrodes. To have improved and better functioning of the supercapacitor, the choice of a separator is crucial. Ionic permeability, porosity, high electrical resistance, low thickness, proper interfacial contact, thermal and chemical stability, and the sustainability of electrolytes are characteristics of a suitable separator. The thickness of the separator is a crucial factor for supercapacitor performance [44]. For instance, bacterial cellulose has a thickness of 1–100 μm, while separators like polyester, polycarbonate, and nylon are in the range of 20–350 μm. Due to their low mechanical strength, a separator thickness less than 1–10 μm would tear easily during the fabrication process. If the consistency is more, there will be a considerable increase in the inter-distance between the electrodes, leading to low capacitance and power density due to the ESR. Porosity is also crucial as pore size matters. Pore size should be greater than the electrolytes' ionic radius for a better flow of ions between the electrodes [45, 46]. Still, it should be smaller than the size of the electrode material to avoid short circuits. Ceramic/glass fibre and polymers or paper act as separators in supercapacitors run with aqueous and ionic electrolytes. Thus, there is a need for an appropriate selection of separators for the proper functioning of the supercapacitor and to attain adequate energy density, power density, self-discharge, and maximum cycle life.

5.3.4 COLLECTOR PLATE

In general, collector plates for both batteries and supercapacitors are inexpensive and highly conductive copper, nickel, iron, and aluminium, ensuring the proper flow of electrons from the active electrode to the external circuit. The connecting link between the collector plates should be highly conductive, have adhesive contact with an electrode material, and have excellent stability during charge/discharge cycles. The appropriate thickness of the collector plate is 20–80 μm. However, due to the low corrosion potential, at high potential, while charging, they may undergo corrosion affecting their contact with the electrode material. Hence, conductive layers like a polymer or titanium or zirconium nitride are inserted between the electrode material and the collector plate. Polymers like polypropylene incorporated with about 50% of fine carbon were also favourable due to their conductivity and stability. Tolerable resistance for a conductive layer is up to 100 Ω/cm [47, 48].

5.3.5 SEALANTS

Sealants play an auxiliary role in avoiding performance loss in the hybrid supercapacitor assembly. They block contaminants like moisture, air, and chemicals, which cause surface oxidation of the electrode and electrolyte degeneracy. Hence, it is necessary to ensure proper sealing or lead shunt resistance between the cells linked to the assembly. Sometimes lead shunt resistance may alter the overall efficiency of the system as well. For a monopolar arrangement, a hermetic seal based on the electrolyte material is ideal for preventing the entry of gas and water. In the case of bipolar structures, shunt current passage is avoided by edge sealing. Shunt current is responsible for the decay in the efficiency of charging and short circuiting. Polymeric materials with high flexibility and excellent moisture resistance are ideal candidates for sealant. Thermosetting epoxy polymer, polyurethane, polyester, and polyacrylate are well-known curable polymers used as sealants to achieve a moisture-resistant and mechanically strong seal in bipolar stacks, preventing short circuits in multiple layers. Rolled cells contain different sealant materials; the internal seal is composed of a curable polymer placed onto the top of the metal casing before it is closed and mechanically crimped shut. To avoid unplanned discharges and create an external barrier against moisture, insulating shrink wrap is used to seal the outer casing electrode. Insulating shrink wrap is made of loosely fitting preformed bags composed of polyolefins (Polypyrrole (PP), Polyethylene (PE), reinforced PE) surrounding the metal casing, followed by heat causing the plastic to shrink and seal around the case insulating the cell. For pouch cells, a multilayer polymer is placed and laminated to bind and seal between the bag layers [49].

5.3.6 DIFFERENT CONFIGURATIONS

As far as a single-cell supercapacitor is concerned, there are three distinct classes of system configurations based the composition of electrode materials: symmetric, asymmetric, and hybrid supercapacitors, even though the distinction between asymmetric and hybrid supercapacitors remains controversial.

5.3.6.1 Symmetric Supercapacitors

Symmetric supercapacitors, as the name suggests, are composed of identical materials on both electrodes, whether it is a carbon-based EDLC electrode or a pseudocapacitive electrode. The majority of commercial devices are asymmetric, based on activated carbon, in combination with an organic electrolyte, with a voltage rating of 2.7 V [50]. This is especially so, owing to the restricted potential window of aqueous electrolytes (1.2 V) which in turn would affect the energy profile of the device. Extensive research is underway to widen the potential window of aqueous electrolytes to so as to exploit the advantages of the same in terms of conductivity, safety, viscosity, dielectric constant, and cost. For instance, Beguin et al. reported an activated carbon-based asymmetric capacitor in 2 M Li_2SO_4 which could reach up to 1.9 V [51]. In a similar study by Hu et al. the device could exhibit a value as high as 2 V [52]. In these studies, the high over-potential due to hydrogen storage mechanism is being explored. In another work by Xia et al., a symmetric pseudocapacitor was fabricated based on RuO_2 which could attain an operating voltage of 1.6 V owing to the high over-potential of the electrodes for oxygen evolution. Hence, a careful selection of electrode and electrolyte materials would be an efficient strategy to improve the potential window and overall performance of symmetric capacitors [53]. It is worth noting that even though both the electrodes are composed of similar materials, the mass loading of the electrodes is different, and it is carefully balanced by considering the ion absorption chemistry at the respective electrodes during the process.

5.3.6.2 Asymmetric Supercapacitors

Asymmetric supercapacitors typically use two different electrodes: a carbon-based EDLC-type electrode and a pseudocapacitive electrode. A combination of activated carbon and MnO_2 is considered to be one of the most extensively studied configurations in this regard [54]. For example, in one such configuration by Brousse et al., a real energy and power density of 10 Wh kg^{-1} and 3.6 kW kg^{-1} could be achieved in 0.65 M K_2SO_4 [55]. A similar design of $AC/K_2SO_4/MnO_2$ by Qu et al. exhibited an energy and power density of 17 Wh kg^{-1} and 2 kW kg^{-1} respectively [56]. A wide range of materials and combinations has been explored in addition to the AC/MnO_2 in terms of materials as well as morphology. CNT/MnO_2, $NiMoO_4/AC$, graphene foam $(GF)/CNTs$, $GF/CNT/PPY$, V_2O_5/electrospun carbon fibre composite, and electrospun carbon fibre [57–60], etc., are a few examples.

5.3.6.3 Hybrid Capacitors

Hybrid capacitors are similar to asymmetric capacitors; however, the pseudocapacitive electrodes here are replaced by battery-type electrodes. Ni- and Co-based electrodes are widely used as the battery-type electrodes. Out of the different battery-type electrodes being investigated, PbO_2 is considered to be one of the best choices owing to its high voltage (2 V) and low cost, especially when combined with the typical activated carbon counter electrode and sulphuric acid electrolyte [54]. According to Yu et al., such a system in 5.3 M H_2SO_4 could deliver a capacitance as high as 71.5 Fg^{-1} with an energy density as high as 32.2 Wh kg^{-1} [61]. However, the overall efficiency of the device is limited by the lower cyclic stability and power performance of the PbO_2 positive electrode. Efforts are under way to improve the performance by various means like nanostructuring and novel electrolyte formulations. $Ni(OH)_2$ is another promising candidate in this regard and a hybrid system with $AC/Ni(OH)_2$ fabricated by Park et al. could attain a specific capacitance as high as 540 Fg^{-1} and a specific energy density of 25 Wh kg^{-1} [62]. There are many such designs in both aqueous and nonaqueous electrolytes reported in the literature, like $AC/Ni(OH)_2$, $AC//LiMn_2O_4$, AC/graphite, etc. [63–65]. In most cases, the device, at lower discharge current densities, shows battery-like behaviour whereas at higher discharge rates they exhibit capacitor-like behaviour. The major challenge associated with these systems is to improve the cyclic stability of the device which is adversely affected by the battery-type electrodes and research is directed in the direction of improving the same.

5.3.7 INTERCONNECTION

A single cell is not enough for large-scale applications, but multiple cells placed in series are required. Due to the slight difference in the capacitance of the manufactured cell, it may lead to leakage of current and resistance, resulting in an imbalance of the cell potential of the supercapacitors stacked in series. Single cells can have capacitive variation of +/− 20%, and hence the overall variation can be much higher than 40%. A proper cell balancing scheme could prevent the above-listed problems and eliminate the imbalance.

There are two techniques of balancing: passive balancing and active balancing. In passive balancing, to compensate for the differences in parallel resistance, a voltage-dividing resistor parallel with each cell reduces current leakage. Passive cell balancing is recommended wherever the cost is to be considered, or the cell is charged/discharged infrequently, and hence they are inefficient. According to some design recommendations, balancing resistors should accommodate current flow 50 times the supercapacitor's worst-case leakage current. Higher resistance resistors can be employed in applications where circuit loss is a significant factor. Although more substantial resistance components offer lower power losses, they slow down the voltage balancing process. The circuit loss and the speed of voltage balancing diminish as resistance rises. The impressed voltage of a supercapacitor limits the most significant resistance value that may be employed. With an increase in impressed voltage, the maximum resistance value drops. Passive voltage balancing is typically employed in systems with a low duty cycle. Supercapacitors used in applications that utilize limited energy sources or higher cycling can operate active voltage balancing circuits since they draw a much lower current steady state. Only when the cell voltage is out of balance are larger currents essential. An operational amplifier or specific Interconnectors (ICs) can be used to implement active balancing control. Although operational amplifiers are more expensive than passive amplifiers, they are often cheaper than a specialized IC. The op-amp and technical IC-based systems have similar convergence rates.

5.4 SUMMARY

Supercapacitors have enormous potential as efficient and green energy storage devices. This chapter discusses essential design considerations like cell voltage, frequency response, lifetime and cycle charging, polarity, heat and temperature effects, and humidity, along with material and manufacturing requirements and implementation methods. Materials used for electrode, electrolyte, separator, and collector plate and the number and arrangement of the single-stacked cells also impact energy storage capabilities. Electrode materials and the design of supercapacitors are the focus of future research to reduce cost, increase energy density, and expand the opportunities in the application field of supercapacitors.

REFERENCES

1. Bu, F., W. Zhou, Y. Xu, Y. Du, C. Guan, & W. Huang. (2020) Recent developments of advanced micro-supercapacitors: Design, fabrication and applications. *Npj Flexible Electronics*, 4(1), 1–16.
2. Muzaffar, A., M.B. Ahamed, K. Deshmukh, & J. Thirumalai. (2019) A review on recent advances in hybrid supercapacitors: Design, fabrication and applications. *Renewable and Sustainable Energy Reviews*, 101, 123–145.
3. Zhi, M., C. Xiang, J. Li, M. Li, & N. Wu. (2013) Nanostructured carbon–metal oxide composite electrodes for supercapacitors: A review. *Nanoscale*, 5(1), 72–88.
4. Han, X., G. Xiao, Y. Wang, X. Chen, G. Duan, Y. Wu, X. Gong, & H. Wang. (2020) Design and fabrication of conductive polymer hydrogels and their applications in flexible supercapacitors. *Journal of Materials Chemistry A*, 8(44), 23059–23095.
5. Wang, G., L. Zhang, & J. Zhang. (2012) A review of electrode materials for electrochemical supercapacitors. *Chemical Society Reviews*, 41(2), 797–828.

6. González, A., E. Goikolea, J.A. Barrena, & R. Mysyk. (2016) Review on supercapacitors: Technologies and materials. *Renewable and Sustainable Energy Reviews*, 58, 1189–1206.

7. Yu, A., V. Chabot, & J. Zhang. (2013) *Electrochemical Supercapacitors for Energy Storage and Delivery: Fundamentals and Applications*. Taylor & Francis.

8. Dai, Z., C. Peng, J.H. Chae, K.C. Ng, & G.Z. Chen. (2015) Cell voltage versus electrode potential range in aqueous supercapacitors. *Scientific Reports*, 5(1), 1–8.

9. Gou, Q., S. Zhao, J. Wang, M. Li, & J. Xue. (2020) Recent advances on boosting the cell voltage of aqueous supercapacitors. *Nano-Micro Letters*, 2(1), 1–22.

10. Lehtimäki, S., A. Railanmaa, J. Keskinen, M. Kujala, S. Tuukkanen, & D. Lupo. (2017) Performance, stability and operation voltage optimization of screen-printed aqueous supercapacitors. *Scientific Reports*, 7(1), 1–9.

11. Zhao, J. & A.F. Burke. (2021) Review on supercapacitors: Technologies and performance evaluation. *Journal of Energy Chemistry*, 59, 276–291.

12. Islam, N., S. Li, G. Ren, Y. Zu, J. Warzywoda, S. Wang, & Z. Fan. (2017) High-frequency electrochemical capacitors based on plasma pyrolyzed bacterial cellulose aerogel for current ripple filtering and pulse energy storage. *Nano Energy*, 40, 107–114.

13. Liu, W., C. Lu, H. Li, R.Y. Tay, L. Sun, X. Wang, W.L. Chow, X. Wang, B.K. Tay, & Z. Chen. (2016) based all-solid-state flexible micro-supercapacitors with ultra-high rate and rapid frequency response capabilities. *Journal of Materials Chemistry A*, 4(10), 3754–3764.

14. Borges, R.S., A.L.M. Reddy, M.-T.F. Rodrigues, H. Gullapalli, K. Balakrishnan, G.G. Silva, & P.M. Ajayan. (2013) Supercapacitor operating at 200 degrees celsius. *Scientific Reports*, 3(1), 1–6.

15. Gualous, H., H. Louahlia-Gualous, R. Gallay, & A. Miraoui. (2009) Supercapacitor thermal modeling and characterization in transient state for industrial applications. *IEEE Transactions on Industry Applications*, 45(3), 1035–1044.

16. Al Sakka, M., H. Gualous, J. Van Mierlo, & H. Culcu. (2009) Thermal modeling and heat management of supercapacitor modules for vehicle applications. *Journal of Power Sources*, 194(2), 581–587.

17. Jiang, D.-e. & J. Wu. (2014) Unusual effects of solvent polarity on capacitance for organic electrolytes in a nanoporous electrode. *Nanoscale*, 6(10), 5545–5550.

18. Murray, D.B. & J.G. Hayes. (2014) Cycle testing of supercapacitors for long-life robust applications. *IEEE Transactions on Power Electronics*, 30(5), 2505–2516.

19. Ng, C.H., H.N. Lim, S. Hayase, Z. Zainal, S. Shafie, H.W. Lee, & N.M. Huang. (2018) Cesium lead halide inorganic-based perovskite-sensitized solar cell for photo-supercapacitor application under high humidity condition. *ACS Applied Energy Materials*, 1(2), 692–699.

20. Anjum, N., N. Joyal, J. Iroegbu, D. Li, & C. Shen. (2021) Humidity-modulated properties of hydrogel polymer electrolytes for flexible supercapacitors. *Journal of Power Sources*, 499, 229962.

21. Ibanez, F., J. Vadillo, J.M. Echeverria, & L. Fontan. Design methodology of a balancing network for supercapacitors. In *IEEE PES ISGT Europe 2013*. 2013. IEEE.

22. Oltean, I., A. Matoi, & E. Helerea. A supercapacitor stack-design and characteristics. in 2010 12th International Conference on Optimization of Electrical and Electronic Equipment. 2010. IEEE.

23. Ibanez, F.M. (2017) Analyzing the need for a balancing system in supercapacitor energy storage systems. *IEEE Transactions on Power Electronics*, 33(3), 2162–2171.

24. Yu, Z., L. Tetard, L. Zhai, & J. Thomas. (2015) Supercapacitor electrode materials: Nanostructures from 0 to 3 dimensions. *Energy & Environmental Science*, 8(3), 702–730.

25. Iro, Z.S., C. Subramani, & S. Dash. (2016) A brief review on electrode materials for supercapacitor. *International Journal of Electrochemical Sciences,* 11(12) 10628–10643.
26. Zhang, L.L. & X. Zhao. (2009) Carbon-based materials as supercapacitor electrodes. *Chemical Society Reviews,* 38(9), 2520–2531.
27. Faraji, S. & F.N. Ani. (2015) The development supercapacitor from activated carbon by electroless plating: A review. *Renewable and Sustainable Energy Reviews,* 42, 823–834.
28. Frackowiak, E., Q. Abbas, & F. Béguin. (2013) Carbon/carbon supercapacitors. *Journal of Energy Chemistry,* 22(2), 226–240.
29. Cai, X., C. Zhang, S. Zhang, Y. Fang, & D. Zou. (2017) Application of carbon fibers to flexible, miniaturized wire/fiber-shaped energy conversion and storage devices. *Journal of Materials Chemistry A,* 5(6), 2444–2459.
30. Wang, X., L. Liu, X. Wang, L. Bai, H. Wu, X. Zhang, L. Yi, & Q. Chen. (2011) Preparation and performances of carbon aerogel microspheres for the application of supercapacitor. *Journal of Solid State Electrochemistry,* 15(4), 643–648.
31. Chen, H., S. Zeng, M. Chen, Y. Zhang, & Q. Li. (2015) Fabrication and functionalization of carbon nanotube films for high-performance flexible supercapacitors. *Carbon.* 92 271–296.
32. Fisher, R.A., M.R. Watt, & W.J. Ready. (2013) Functionalized carbon nanotube supercapacitor electrodes: A review on pseudocapacitive materials. *ECS Journal of Solid State Science and Technology,* 2(10), M3170.
33. Abdah, M.A.A.M., N.H.N. Azman, S. Kulandaivalu, & Y. Sulaiman. (2020) Review of the use of transition-metal-oxide and conducting polymer-based fibres for high-performance supercapacitors. *Materials & Design,* 186, 108199.
34. Ma, Y., X. Xie, W. Yang, Z. Yu, X. Sun, Y. Zhang, X. Yang, H. Kimura, C. Hou, & Z. Guo. (2021) Recent advances in transition metal oxides with different dimensions as electrodes for high-performance supercapacitors. *Advanced Composites and Hybrid Materials,* 4(4), 1–19.
35. Zheng, M., X. Xiao, L. Li, P. Gu, X. Dai, H. Tang, Q. Hu, H. Xue, & H. Pang. (2018) Hierarchically nanostructured transition metal oxides for supercapacitors. *Science China Materials,* 61(2), 185–209.
36. Delbari, S.A., L.S. Ghadimi, R. Hadi, S. Farhoudian, M. Nedaei, A. Babapoor, A.S. Namini, Q. Van Le, M. Shokouhimehr, & M.S. Asl. (2021) Transition metal oxide-based electrode materials for flexible supercapacitors: A review. *Journal of Alloys and Compounds,* 857, 158281.
37. Snook, G.A., P. Kao, & A.S. Best. (2011) Conducting-polymer-based supercapacitor devices and electrodes. *Journal of Power Sources,* 196(1), 1–12.
38. Meng, Q., K. Cai, Y. Chen, & L. Chen. (2017) Research progress on conducting polymer based supercapacitor electrode materials. *Nano Energy,* 36, 268–285.
39. Béguin, F., V. Presser, A. Balducci, & E. Frackowiak. (2014) Carbons and electrolytes for advanced supercapacitors. *Advanced Materials,* 26(14) 2219–2251.
40. Lewandowski, A., A. Olejniczak, M. Galinski, & I. Stepniak. (2010) Performance of carbon–carbon supercapacitors based on organic, aqueous and ionic liquid electrolytes. *Journal of Power Sources,* 195(17), 5814–5819.
41. Cheng, Y., H. Zhang, S. Lu, C.V. Varanasi, & J. Liu. (2013) Flexible asymmetric supercapacitors with high energy and high power density in aqueous electrolytes. *Nanoscale,* 5(3), 1067–1073.
42. Pan, S., M. Yao, J. Zhang, B. Li, C. Xing, X. Song, P. Su, & H. Zhang. (2020) Recognition of ionic liquids as high-voltage electrolytes for supercapacitors. *Frontiers in Chemistry,* 8, 261.
43. Zhang, L., S. Yang, J. Chang, D. Zhao, J. Wang, C. Yang, & B. Cao. (2020) A review of redox electrolytes for supercapacitors. *Frontiers in Chemistry,* 8, 413.

44. Verma, K.D., P. Sinha, S. Banerjee, K.K. Kar, & M.K. Ghorai. (2020) Characteristics of separator materials for supercapacitors. In *Handbook of Nanocomposite Supercapacitor Materials Ispringer*, p. 315–326.

45. Sun, X.-Z., X. Zhang, B. Huang, & Y.-W. Ma. (2014) Effects of separator on the electrochemical performance of electrical double-layer capacitor and hybrid battery-supercapacitor. *Acta Physico-Chimica Sinica*, 30(3), 485–491.

46. Deka, B.K., A. Hazarika, J. Kim, Y.B. Park, & H.W. Park. (2017) Recent development and challenges of multifunctional structural supercapacitors for automotive industries. *International Journal of Energy Research*, 41(10), 1397–1411.

47. Liu, L., H. Zhao, Y. Wang, Y. Fang, J. Xie, & Y. Lei. (2018) Evaluating the role of nanostructured current collectors in energy storage capability of supercapacitor electrodes with thick electroactive materials layers. *Advanced Functional Materials*, 28(6), 1705107.

48. Arvani, M., J. Keskinen, D. Lupo, & M. Honkanen. (2020) Current collectors for low resistance aqueous flexible printed supercapacitors. *Journal of Energy Storage*, 29, 101384.

49. Wu, Z.-S., X. Feng, & H.-M. Cheng. (2014) Recent advances in graphene-based planar micro-supercapacitors for on-chip energy storage. *National Science Review*, 1(2), 277–292.

50. Burke, A. (2007) R&D considerations for the performance and application of electrochemical capacitors. *Electrochimica Acta*, 53(3), 1083–1091.

51. Demarconnay, L., Raymundo-Piñero, E., & Béguin, F. (2010) A symmetric carbon/carbon supercapacitor operating at 1.6 V by using a neutral aqueous solution. *Electrochemistry Communications*, 12(10), 1275–1278.

52. Wu, T.H., Hsu, C.T., Hu, C.C., & Hardwick, L.J. (2013) Important parameters affecting the cell voltage of aqueous electrical double-layer capacitors. *Journal of Power Sources*, 242, 289–298.

53. Xia, H., Meng, Y.S., Yuan, G., Cui, C., & Lu, L. (2012) A symmetric RuO_2/RuO_2 supercapacitor operating at 1.6 V by using a neutral aqueous electrolyte. *Electrochemical and Solid State Letters*, 15(4), A60.

54. Wang, Y., Song, Y., & Xia, Y. (2016) Electrochemical capacitors: Mechanism, materials, systems, characterization and applications. *Chemical Society Reviews*, 45(21), 5925–5950.

55. Brousse, T., Toupin, M., & Belanger, D. (2004) A hybrid activated carbon-manganese dioxide capacitor using a mild aqueous electrolyte. *Journal of the Electrochemical Society*, 151(4), A614.

56. Qu, Q., Zhang, P., Wang, B., Chen, Y., Tian, S., Wu, Y., & Holze, R. (2009) Electrochemical performance of MnO_2 nanorods in neutral aqueous electrolytes as a cathode for asymmetric supercapacitors. *The Journal of Physical Chemistry C*, 113(31), 14020–14027.

57. Ou, T.M., Hsu, C.T., & Hu, C.C. (2015) Synthesis and characterization of sodium-doped MnO_2 for the aqueous asymmetric supercapacitor application. *Journal of the Electrochemical Society*, 162(5), A5124.

58. Li, L., Peng, S., Wu, H.B., Yu, L., Madhavi, S., & Lou, X.W. (2015) A flexible quasi-solid-state asymmetric electrochemical capacitor based on hierarchical porous V_2O_5 nanosheets on carbon nanofibers. *Advanced Energy Materials*, 5(17), 1500753.

59. Peng, S., Li, L., Wu, H.B., Madhavi, S., & Lou, X.W. (2015) Controlled growth of $NiMoO_4$ nanosheet and nanorod arrays on various conductive substrates as advanced electrodes for asymmetric supercapacitors. *Advanced Energy Materials*, 5(2), 1401172.

60. Liu, J., Zhang, L., Wu, H.B., Lin, J., Shen, Z., & Lou, X.W.D. (2014) High-performance flexible asymmetric supercapacitors based on a new graphene foam/carbon nanotube hybrid film. *Energy & Environmental Science*, 7(11), 3709–3719.
61. Yu, N., & Gao, L. (2009) Electrodeposited PbO_2 thin film on Ti electrode for application in hybrid supercapacitor. *Electrochemistry Communications*, 11(1), 220–222.
62. Yin, J.L., & Park, J.Y. (2015) Asymmetric supercapacitors based on the in situ-grown mesoporous nickel oxide and activated carbon. *Journal of Solid State Electrochemistry*, 19(8), 2391–2398.
63. Amatucci, G.G., Badway, F., Du Pasquier, A., & Zheng, T. (2001) An asymmetric hybrid nonaqueous energy storage cell. *Journal of the Electrochemical Society*, 148(8), A930.
64. Du Pasquier, A., Plitz, I., Gural, J., Menocal, S., & Amatucci, G. (2003) Characteristics and performance of 500 F asymmetric hybrid advanced supercapacitor prototypes. *Journal of Power Sources*, 113(1), 62–71.
65. Cheng, L., Liu, H.J., Zhang, J.J., Xiong, H.M., & Xia, Y.Y. (2006) Nanosized $Li_4Ti_5O_{12}$ prepared by molten salt method as an electrode material for hybrid electrochemical supercapacitors. *Journal of the Electrochemical Society*, 153(8), A1472.

6 Characterization Techniques

Characterization

Techniques

*Deepthi Panoth, Kunnambeth M. Thulasi,
Fabeena Jahan, Sindhu Thalappan Manikkoth,
Divya Puthussery,
Baiju Kizhakkekilikoodayil Vijayan,
and Anjali Paravannoor*

CONTENTS

6.1 INTRODUCTION

Electrochemical energy storage technologies presently exhibit a pivotal and promising role in addressing energy-related issues globally [1]. Globally, energy security has become the highest priority among individuals and governments. Over the past few decades, researchers and scientists have been focusing on the development of supercapacitors with high energy and power density, high-surface-area novel electrode materials [2]. From time to time, researchers have made several attempts to recognize, tackle, and eliminate the uncertainties caused by the traditionally used instruments, calculation methods, major influencing performance factors, and parameters closely related to the performance evaluation of supercapacitors. A variety of techniques has been employed to measure and evaluate the characteristic parameters of supercapacitors with high precision and accuracy, both in the research and industrial fields. Some of the major parameters for evaluating the performance of supercapacitors are operating voltage, power density, cell capacitance, energy density, equivalent series resistance (ESR), and time constant. Standard test procedures are necessary to evaluate detailed and reliable information regarding the electrochemical performance of supercapacitors. The common characterization techniques used to measure the performance of supercapacitors includes cyclic voltammetry, galvanostatic charge/ discharge curves, and electrochemical impedance spectroscopy.

Cyclic voltammetry (CV) is an electrochemical technique that helps us to investigate information such as the charge-transfer kinetics of electrodes, the formal potential, and its strong correlation with the standard reduction potential of redox couples, the diffusion coefficient of redox couples, and chemical reaction mechanism [3]. Galvanostatic charging/discharging or the chronopotentiometry technique has been extensively used to evaluate and characterize the electrochemical performance of supercapacitors owing to its comparably direct physical relationship to the capacitive charge. In the case of an ideal capacitor, with constant current a triangular response with a slight deviation at the vertex due to the solution resistance is predictable. This method also helps us to evaluate the longevity of the electrode materials and the cyclic stability of the supercapacitors [4].

Electrochemical impedance spectroscopy (EIS) is an extremely sensitive characterization technique used to evaluate the capacitance of energy storage devices [5]. EIS measurements comprise an electric potential via a specific frequency about the working electrode. Electrochemical impedance is the ratio of amplitudes of oscillating potential and oscillating current and is measured as a potential drop [6]. The modelling technique is a mathematical method used to evaluate and predict the performance of supercapacitors. Currently, modelling is crucial for the design and prediction of performance and lifetime of supercapacitors. The cell parameters, design constraints, or electronic device characteristics have a profound impact on the current and potential measurements, which affects the accuracy of measurements. The inaccuracy of measurement using these techniques remains a source of controversy. This chapter summarizes the different characterization techniques used to evaluate the electrochemical performance of supercapacitors.

6.2   CYCLIC VOLTAMMETRY (CV)

The basic electrochemical characterization method for materials is CV [3, 7, 8]. Owing to its versatility, it is widely exploited in the electrochemical analysis of materials. It has been properly employed in laboratory-scale testing as it enables qualitative/quantitative/kinetic studies by scanning over a broad range of scan rates. However, it is technically difficult to use these techniques in commercial-scale test cells, as it leads to ominously large magnitudes of current. In this technique, the current is recorded by sweeping the voltage back and forth between the chosen limits. Many significant inferences regarding the material and its capacitive properties, the reversible/irreversible/quasireversible behaviour of the electrode system, etc., can be properly achieved by the analysis of the cyclic voltammograms obtained.

In CV operation, the electrical voltage is linearly varied with time between the reference and working electrodes for three electrode systems, while in two electrode systems the voltage is linearly varied between the positive and negative electrodes. The integration of the electric current with respect to time gives the amount of total charge accumulated at the surface; the total charge divided by the voltage window gives the capacitance. The range of voltage change is called the voltage window and the rate of voltage change is called the scan rate or sweep rate. During operation, the current is simultaneously recorded with cathodic and anodic sweepings, in order to characterize the corresponding electrochemical reactions. The results are plotted as voltage vs current or sometimes voltage vs time. Thus, in simple terms, we can say that the application of a potential ramp to a device or an electrode between two potential limits and the measurement of the resulting current describes the basic principle of CV.

The electrochemical reactions during the charging and discharging process of a supercapacitor can be easily deduced from the shape of CV curves. When a super-capacitor is charged from zero potential, an initial increase in current occurs, which decreases further upon increasing the potential.

An ideal double-layer supercapacitor with a slight resistance shows a perfect rect-angular shape for the CV curve and is given in Figure 6.1. Scan rate-dependent behaviour is often exhibited by ideal supercapacitors. However, it is impossible for an ideal capacitor that addresses all concerns to exist. The real supercapacitors suf-fer from various limitations and imperfections and thus deviate from the expected regular rectangular shape as given in Figure 6.1. Thus, a real capacitor is represented as overall capacitance C and internal resistance R combined in series. A proper volt-age window is there for the proper working of a supercapacitor; voltages outside the window damage the device through electrolyte decomposition. By continuous

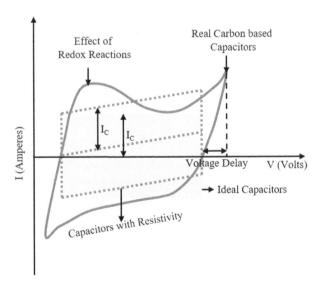

FIGURE 6.1   Schematic diagram showing cyclic voltammetry characteristics of superca-pacitors [9].

adjustment of the voltage window in a three-electrode system, it is easy to determine the suitable operating voltage window for supercapacitive materials. There are aqueous and non-aqueous electrolytes for supercapacitors. Though aqueous electrolytes are normally harmless and easy to handle, supercapacitors with aqueous electrolytes suffer from a narrow potential window in comparison with non-aqueous electrolytes.

The integration of the CV curves helps in the calculation of specific capacitance and energy performance of the supercapacitive materials [10]. Thus, the capacitance (C) values in units Fcm^{-2} can be calculated from the CV curves by using Equation (6.1).

$$C = \frac{1}{Av(V_2 - V_1)} \int_{V_1}^{V_2} I(V) dV \qquad (6.1)$$

where A is the area of the working electrode in cm^2, v is the scan rate in Vs^{-1} and (V_2–V_1) is the potential window expressed in V. In electrochemical characterizations, capacitance is usually measured at altered scan rates. The capacitance values obtained at higher scan rates are lower than the same at lower scan rates, owing to the reduced diffusion of the electrolyte ions at higher scan rates.

The equal process can also be utilized for supercapacitor devices to calculate the total cell capacitance.

The energy stored in a supercapacitor E is directly proportional to its capacitance C and is given by Equation (6.2).

$$E = \frac{1}{2} CV^2 \qquad (6.2)$$

The energy utilized per unit of time gives the power. The internal components of the capacitor contribute to a resistance, which is calculated as the ESR and thus the maximum power, P_{max} for a supercapacitor can be calculated by Equation (6.3).

$$P_{max} = \frac{V^2}{4 \times ESR} \qquad (6.3)$$

The CV technique utilizing a three-electrode configuration is considered as the most promising one to determine the charge storage mechanisms of supercapacitive materials. The shape of CV curves helps in the prediction of capacitive behaviour of the electrodes, whether it is EDLC or pseudocapacitive. For EDLCs and most pseudocapacitive materials, the shape of CV curves is rather rectangular. Pronounced redox peaks in a highly reversible fashion may occur in the case of some pseudocapacitor materials. Therefore, we cannot discriminate between the materials as EDLCs or pseudocapacitive merely by the shape of CV curves. A more quantitative and trustworthy method is developed by Trasatti for deducing the contributions from EDL and pseudocapacitive mechanisms separately from the CV data [11, 12]. A linear relationship exists between the reciprocal of capacitance (C^{-1}) and the square root of scan rate ($v^{1/2}$), as given by Equation (6.4).

$$C^{-1} = constant * v^{\frac{1}{2}} + C_T^{-1} \tag{6.4}$$

Here C, C_T, and v represent calculated capacitance, total capacitance, and scan rate respectively. Thus, the reciprocal of the y-intercept of the C^{-1} vs $v^{1/2}$ plot equals the total capacitance, which is the sum of pseudocapacitance and electrical double-layer capacitance.

The calculated capacitance (C) and the reciprocal of the square root of scan rate ($v^{-1/2}$) are also linearly related by Equation (6.5).

$$C = constant * v^{-\frac{1}{2}} + C_{EDL} \tag{6.5}$$

Here C, C_{EDL}, and v are the calculated capacitance, maximum electrical double-layer capacitance, and scan rate, respectively. The y-intercept of the C vs $v^{-1/2}$ plot equals the maximum electrical double-layer capacitance. The awareness of total capacitance and maximum electrical double-layer capacitance helps in the calculation of maximum pseudocapacitance. The percentage capacitance contribution can be calculated by Equations (6.6) and (6.7).

$$C_{EDL}\% = \frac{C_{EDL}}{C} \times 100\% \tag{6.6}$$

$$C_P\% = \frac{C_P}{C} \times 100\% \tag{6.7}$$

Although CV is helpful in the evaluation of cyclability of an electrode or a supercapacitor device, galvanostatic charge/discharge cycling is favoured while executing such studies.

6.3 GALVANOSTATIC CHARGE/DISCHARGE OR CHRONOPOTENTIOMETRY

The galvanostatic charge/discharge (GCD) technique is a reliable strategy used to determine the electrochemical capacitance of electrode materials where continuous charging and discharging is performed for a supercapacitor device or a single electrode (in a three-electrode system). A GCD test helps us to determine the durability of the electrode material, or it gives us the stability of the electrochemical storage devices. GCD uses a constant current to charge or discharge the device. The charge/discharge cycles are recorded as a function of time with varying potential. The charging and discharging rates affect various parameters such as capacitance, resistance, and cyclability. This technique is also called chronopotentiometry.

The voltage variation with respect to time is given by the following equation:

$$V(t) = Ri + \frac{t}{C}i(V)$$ (6.8)

where Ri is the series resistance and C is the capacitance of the supercapacitor, V is the voltage window and t is the discharge time. The capacitance, power density, and energy density of the supercapacitor device can be calculated from the discharge curve. The slope of the charge/discharge curves of chronopotentiometry gives the capacitance of the supercapacitor and this can be calculated from the GCD curves using Equation 6.10.

$$C = I\frac{\partial t}{\partial V}(F)$$ (6.9)

$$C = I\frac{" t}{" V}(F)$$ (6.10)

where I is the set current, V is the voltage window, Δt is the discharge time and C is the capacitance in farads. From the voltage drop (V_{drop}) across the current inversion (I), the series resistance can be derived.

$$R = \frac{V_{drop}}{\Delta I}(\Omega)$$ (6.11)

In this technique, the shape of the resultant curve delivers relevant information concerning the charge storage mechanism, percentage contribution to faradaic and non-faradaic charge storage, and effective series resistance of the supercapacitor. Here a Ragone plot is constructed by repeating the charge discharge process for different discharge times [4].

In EDLCs, when a constant current is applied the potential varies linearly with time and thus a linear plot is obtained as a result of interfacial charging between the electrode interface and the electrolyte during the non-faradaic process [4]. Figure 6.2(a) shows the schematic diagram of a typical charge/discharge curve or galvanostatic curve on an EDLC. An instantaneous voltage jump is observed by applying a constant current. The presence of series resistance initiates a voltage drop via the ohmic current flow in supercapacitors. With time, the voltage increases linearly until the preset voltage boundary, preceded by a sudden voltage drop owing to the disappearance of the current is described as IR drop. Due to self-discharge redistribution, the voltage may drop during the rest time between charging and discharging. At higher current densities this redistribution of charge phenomenon is found to be more dominant. When the supercapacitors are fully charged, the remaining ions try to diffuse into the unoccupied areas on the surface of the electrode. This diffusion-controlled process continues until the charges are uniformly distributed all over the surface of the electrode. Normally this type of self-discharge occurs soon after charging is finished and this lasts for a while based on the structural properties of

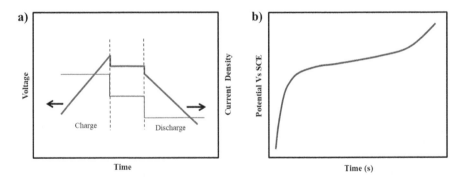

FIGURE 6.2 (a) Typical charge/discharge galvanostatic charge discharge curve and (b) choropotentiogram for charging process with faradaic reaction.

the electrode materials. Consequently, a slow voltage drop is witnessed when charge redistribution takes place.

Unlike in EDLC, pseudocapacitors or hybrid capacitors exhibit a non-linear nature for the voltage vs time graph due to the faradaic reactions. In general, during charging, the voltage increases linearly at the initial stage of charging in a chronopotentiogram. The increase in voltage decelerates as a result of the faradaic reactions at the electrode/electrolyte interface until the reactants in the electrolytes are depleted and the diffusion of reactants is no longer adequate to maintain the reaction. Thus, in this range, the chronopotentiogram exhibits nonlinearity (Figure 6.2(b)). It shows a non-linear chronopotentiogram for a pseudocapacitor and the capacitance from a non-linear V-t curve can be calculated by integrating the inverse slope, $\Delta t/\Delta V$ as shown in the following equation:

$$C = I \int \frac{dt}{V(t)}$$ (6.12)

where I is the constant current, ΔV is the potential difference, and Δt is the time for the voltage changes.

The specific energy of supercapacitor devices can be calculated using both CV and GCD techniques [3]. The chronopotentiogram of EDLC and pseudocapacitors displays a nearly linear charge/discharge curve making a symmetric triangle. During measurement a constant current is applied, the specific energy can be determined from the following mathematical expression given below from the GCD technique.

$$E_S = \int_0^Q V_0 dq = \frac{1}{2} V_0 Q = \frac{1}{2} V_0 I t_c$$ (6.13)

where V_0 is the peak potential value, Q is the net charge obtained by the device by applying a constant current over a charging time of t_c. Thus, specific energy can be calculated after measuring these parameters from the GCD technique. But for non-linear chronopotentiogram, Equation (6.13) cannot give the specific energy of the supercapacitor device as it is difficult to get simple solutions for integration as the current varies non-linearly [13].

6.4 ELECTROCHEMICAL IMPEDANCE SPECTROSCOPY (EIS)

EIS is considered to be a valid technique for the electrochemical characterization of power sources and energy storage materials. The impedance method was first developed from the conventional dielectric systems in the early 1950s [14]. In the 1980s, frequency response analyzers and AC impedance spectroscopy became a familiar tool for the electrochemical characterization of energy storage systems [15].

The evaluation of capacitor behaviour is mainly based on the frequency-response characteristics, quantification of ESR, and modelling of equivalent circuits and in turn depend upon various parameters such as the intrinsic nature, surface area, pore-size distribution, and thickness of the active material and also the nature of electrolyte. EIS helps to measure the frequency responses and power limiting internal resistance with great accuracy. Moreover, as the electrochemical processes vary in accordance with their own time constant, the measuring technique should cover a wide range of time and the EIS method enables the same as it covers a large time range from microseconds to hours. Additionally, the method gives high precision values as it receives sufficient acquisition times as the measurements are established under steady-state conditions [16].

In the EIS technique, the perturbation of the system is accomplished by applying a small amplitude of the alternating signal of about 5–10 mV to get a linear response and observing a return of the system to a stationary state. The technique can be performed either by controlling the current or the voltage and measuring either voltage or current. In most cases, the voltage control technique is employed, however, the same conclusion can be obtained by the current control method.

The voltage control technique is accomplished by setting the voltage of the electrochemical system (Vs) and a sinusoidal signal of small amplitude is applied, and the measurement is monitored at different frequencies ($f = \omega / 2\pi$). δv and δi are the amplitudes of the voltage and the current, respectively. Since the signal is very small, a linear relationship can be derived between current and voltage for each pulse applied (Equation (6.14)).

$$V = Zi(V) \tag{6.14}$$

where V is the voltage, i is the current, and Z is the impedance.

In the case of an ideal electrochemical cell with a particular amplitude and frequency of applied voltage, the cell gives an alternating current response at the same frequency. However, in real systems, the current response may be complex owing to the contributions of other frequencies arising from the components. Furthermore, the frequencies of the applied perturbation signal are also prone to a deviation between the mHz and kHz ranges. Thus, a shift in amplitude and time between input and output signal with applied frequency can be observed. These factors rely on the rates of physical process in the electrochemical cell and the response of the system towards the oscillating signal. As there are different electrochemical processes occurring in a system, different frequencies can separate the processes with different time scales. For instance, at higher frequencies, a change in the direction of the applied field occurs much faster than the chemistry response, thus capacitance contribution at higher frequency was dominated by the charge/discharge process of the electrical double layer. However, at lower frequencies, slow electrochemical reactions such as diffusion dominate with the alternating signal [17].

As we discussed earlier, in real systems, the resulting impedance formed from the superposition of a sinusoidal signal and observing the magnitude and phase angle of the resulting AC current was represented as a complex quantity as in Equation (6.15).

$$Z(\omega) = Z'(\omega) + jZ''(\omega) \qquad (6.15)$$

where Z' and Z'' are the real and imaginary parts of $Z(\omega)$, respectively.

From the equation, it is evident that the EIS technique is suitable for the linearization of complex electrochemical systems as well. The impedance measurements also aid in modelling an analogous electric circuit known as the equivalent circuit that portrays the behaviour of the electrochemical cell that has been studied. The circuit represents the fingerprint of the active material and provides much valuable information about the physical and chemical properties of the active material, electrode/electrolyte interface, and reaction kinetics. As there are many equivalent circuits that can fit the experiment data, we must take care to select a particular circuit that matches the electrochemistry of the system. Figure 6.3(c) denotes the Randles equivalent circuit which is most commonly used to explain simple electrochemical systems. The circuit includes a series resistance (R_Ω) related to bulk electrolyte resistance, a double-layer capacitance (C_{dl}) related to the accumulation of charge at the electrode/electrolyte interface, and charge-transfer resistance (R_{ct}) related to the current exchanged which defined using Butler–Volmer equation. The charge-transfer resistance is in parallel with the double-layer capacitance. The component Z_w is the Warburg impedance which represents the polarization of electrochemical cells caused by diffusion limitation [15]. The Randles circuit also serves as a building block for more complex circuit models. For instance, Masarapu et al. used a modified Randles circuit to study the capacitance of carbon nanotube supercapacitors [18]. The examination of impedance by EIS spectra is done by using either Nyquist or Bode plots. As mentioned earlier, the impedance $Z(\omega)$ consists of a real part and an imaginary part. The Nyquist plot is obtained by plotting the real part on the x-axis and the imaginary part on the y-axis. Here the y-axis is negative and each point in the plot gives the value of impedance at one particular frequency. The impedance Nyquist plots of an electrochemical system show two different frequency-dependent regions consisting of a high frequency region on the left followed by a low frequency region on the right.

In EDLCs, the high frequency region is comprised of a semicircle and low frequency region is comprised of a vertical line. The semi-circular region corresponds to the charge-transfer reactions of the electrochemical system and is attributed to the double-layer capacitance parallel to the charge-transfer resistance (R_{ct}) at the

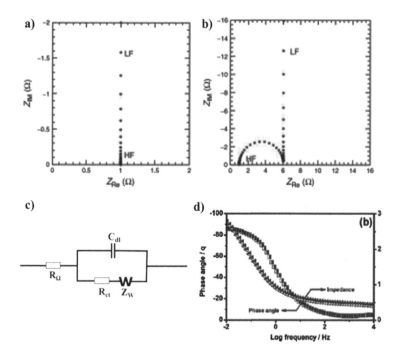

FIGURE 6.3 (a) Nyquist plot of an ideal electrical double-layer capacitor and (b) Nyquist plot of an ideal pseudocapacitor [16]. (c) Schematic illustration of Randles equivalent circuit and (d) Bode plot of an ideal pseudocapacitor.

electrode-electrolyte interface and is calculated as the diameter of the semicircle. The series resistance (R_Ω) which is accounted as the sum of internal resistance of the active material, electrolyte resistance and the contact resistance at the solid/liquid interface determined as the x-axis intercept at high frequency region.

The linear part in the low frequency region is clearly related to the diffusion resistance/Warburg resistance of the electrolyte into the interior of the electron surface and ion diffusion into the electrode surface [19]. At higher frequencies of the applied potential, the Warburg impedance has a very small value as the reactants can't move to longer distances. However, at lower frequencies, the diffusing reactants move further, thereby increasing the Warburg impedance. The Nyquist plot for an ideal electrical double-layer and pseudocapacitor is shown in Figure 6.3. Figure 6.3(a) represents the behaviour of an EDLC, which can be described by a series resistance (R_Ω) and a capacitance (C_{dl}). Figure 6.3(b) represents the behaviour of the pseudocapacitor, which is explained by electrolyte resistance and double-layer capacitance in parallel with the charge transfer resistance involved in the faradaic process. In this case, the capacitance is related to the accumulation of charge at the electrode/electrolyte interface which is different from pure EDLCs. Here the charge transfer resistance exists, and mass transfer limitation could also appear. The Bode plot consists of two plots in a single plot. The logarithmic scale of the frequency is plotted on the x-axis. Logarithms of the impedance on one y-axis and phase shift φ on the second y-axis. A Bode plot gives a clear idea about the capacitor in parallel to a resistor as the peak in the phase shift. Unlike the Nyquist plot, the Bode plot doesn't show frequency information (Figure 6.3(d)).

The important advantage of the EIS technique is the use of very small AC signals to investigate the electrical characteristics without disturbing the properties of the system. However, as a comparison with other techniques, the resistance value obtained from the EIS technique is smaller than the resistance value obtained from the constant current charge/discharge technique and thus limited in explaining the power performance of supercapacitor devices. However, EIS can be considered as a potential technique to determine the impedance, charge transfer, specific capacitance, mass transport, and the mechanism of charge storage in supercapacitor devices using a three-electrode system.

6.5 MODELLING TECHNIQUES

Modelling an electrochemical system like a supercapacitor refers to the mathematical representation of various aspects of supercapacitor performance for describing and analyzing the system. It is always a prerequisite for researchers and engineers in order to fulfil a specific purpose regarding their assumptions and requirements that will help in the improvement of materials and processes employed for supercapacitor manufacture [20]. As the experimental approaches are usually time consuming and expensive, physical modelling could ease the optimization of electrode morphology as well as the identification of the best electrolyte in a more systematic approach. Thus, modelling is essential for the prediction of design, performance, lifetime, cost, etc., of a supercapacitor for its control synthesis. Modelling is important to implement the electrochemical principles and to optimize the incorporation and management of a supercapacitor in an entire electrical system. As such, many models are reported in various works for various purposes, like capturing electrical and thermal activities, self-discharge, simulation of ageing, etc. A review of the main modelling progress is discussed in this chapter identifying the core strengths and weaknesses as well as the most sensible properties of each model [21, 22].

Basically, three supercapacitor modelling approaches are in common practice: a mathematical or electrochemical model, an electric circuit or equivalent circuit model, and a non-electric circuit model, each having their own advantages and disadvantages. The mathematical modelling approach is highly precise but possesses complicated computations and demands a lot of parameters that need to be experimentally recognized as they define real reactions inside the supercapacitors at the cost of coupled partial differential equations. Moreover, it does not usually have a clear physical meaning and cannot easily be incorporated into a circuit diagram and thus hinders their use in real-time energy control and management in the whole system. The non-electric circuit models also have the same limitations as the mathematical modelling. In contrast, electric circuit models are simpler to implement and need only low computational efforts but can define the behaviour of supercapacitors in comparatively lesser condition criteria making them a real-time energy management synthesis. They are derived from empirical and experimental data subject to certain conditions and make them insufficient for expressing the supercapacitor dynamics under a wide range of conditions, causing model mismatch problems. Also, they are not able to incorporate all the parameters and cannot study different physical representations simultaneously, so no internal information will be available. To have a more accurate and ideal model, it is advised to develop modern theories for supercapacitors that can support the experiments and are achieved with the help of molecular dynamics and advanced computing software [22–24].

In this chapter, we reviewed the progress developed so far in the electrochemical modelling of supercapacitors. Novel characterization techniques develop complex signal outputs whose relation to the fundamental structure or physical phenomena benefits from computational modelling. It is expected that the novel spectroscopic methods give a boost to the theoretical approaches to the limit and further development of more advanced techniques are required that treat more precisely the dynamic interaction between electrons and between electrons and rapidly varying external potentials. The following classes of models are proposed by researchers for an electrochemical supercapacitor: empirical models, dissipative transmission line models, continuum models, atomistic models, quantum models, and simplified analytical models [25]. Each modelling approach was developed generally based on the equivalent-circuit approach or porous electrode theory and for a distinct purpose and, thus, displays various advantages and limitations [26].

6.5.1 EMPIRICAL MODELLING

An empirical electrical equivalent model is considered to have three branches and the immediate branch capacitance is assumed to be voltage dependent whereas the long-term branch accounts for a maximum 30 minutes charge variation. Empirical models offer a better description of the electrical characteristics than a simple RC circuit. In addition, it enables the engineers to easily incorporate the model into the simulation system and it could also be used to characterize parameters such as self-discharge, diffusion controlled self-discharge, and current leakage. Experimental records are simulated to obtain specific capacitance and ESR values which in turn can be used to calculate the energy and power density. The simulated parameter values obtained considering both voltage-independent parallel leakages and solvent decomposition could be utilized for the thorough understanding of supercapacitors charging and discharging behaviours as well as the performance/efficiency losses through experimental validation. Limiting workable cell voltage can also be calculated with these models [27].

Zubieta and Bonert reported the first empirical models that addressed the rate of change of capacitance or terminal voltage. Simulated results were in good agreement with the experimental data for operating voltages greater than 1 V and when voltage is decreased, the error is found to be increased for the same charge cycles [28]. These models are very useful but have setbacks as well as they are not derived from the fundamental basis of physics. For the operating condition that differs from those intended for parameter extraction and for the longer time intervals where the internal temperature rise alters the electrical characteristics of the supercapacitor, the results from empirical modelling seem inaccurate. Thus, these models cannot be employed to calculate the lifespan of a particular supercapacitor [29].

6.5.2 DISSIPATION TRANSMISSION LINE MODELS

The dissipative transmission line model's main advantage is that it offers a direct correlation between pore structure and time response and imparts a first-order approximation of the exponential decay/rise in the voltage level assuming a constant current charge/discharge cycle. These are built based on the physical structure of the electrode/electrolyte interface, thus are utilized mainly for purposes like electrode synthesis improvement, prediction of the electrode's surface impedance, the study of self-discharge behaviour, etc. It is also helpful to find the relationship between electric and physical characteristics, as well as to simulate the frequency response and relaxation time dependency on several parameters of a supercapacitor. The electrode's porous structure which restricts the ion movement to interfacial sites situated within a pore is the central reason for the nonlinear rise of terminal voltage [30, 31]. Thus, one must have a model that explains the steady-state and transient response of the porous electrode region. The transmission line considers the apt representation of the electric circuit for a supercapacitor. The porous electrode model was first created by de Levie. He considered the interfacial capacitance of an electrode wall with pores as a dissipative transmission line and as a distributed RC circuit. He defines the variation of current, I and voltage, V with respect to position x and time t along the pore by the following Equations (6.16) and (6.17) [32].

$$\frac{dV}{dx} = -\text{IR} \qquad\qquad (6.16)$$

$$\frac{dI}{dx} = -C\frac{dV}{dt} \qquad\qquad (6.17)$$

where R and C are resistance and capacitance per unit length, respectively. De Levie presumed that the solution concentration and hence the capacitance is constant all through the pore. Usually, capacitance is found to be maximum at low frequency and minimum at higher frequency. In conclusion, for a given voltage, current and energy are found to be relatively larger than that predicted by classical experiments in the case of constant capacitance. The performance of the supercapacitor is analyzed with an impedance spectrum analyzer [33]. To explain the dependences of temperature, voltage, and frequency, Rafik et al. produced a simple equivalent electrical circuit using a combination of the Zubieta voltage model and De Levie frequency model with an extra circuit that accounts for the temperature dependence of supercapacitors at low frequency. It was figured out that temperature has zero effect on capacitance at lower frequencies [27].

This modelling approach considers only a low number of interfacial dynamics and their major disadvantage is that it is challenging to measure capacitance when number of cells are connected in series or parallel combination.

6.5.3 CONTINUUM MODELS

To represent the electrolyte/electrode interface, the continuum modelling approach exploited the use of the Poisson–Nernst–Planck concept of electro diffusion, and the physicochemical parameters were used as functions of their vicinity owing to the high electric field at the interface instead of keeping it a constant [34]. Based on continuum theory a 3D model was developed by Hainan Wang and Laurent Pilon for the simulation of electrode double-layer supercapacitance with an electrode with ordered mesoporous structures. They set a new boundary condition for the Stern layer without simulating it in the electrolyte regime and it is then used to simulate the morphology of electrode CP204-S15 carbon with mesoporous, manufactured and characterized by Woo et al. [35, 36]. The continuum model is also used to study the combined effects of diffusion coefficient for intercalation, scan rate, and electrode thickness. It also accounts for the variation of pseudocapacitance along with the electrode thickness [34].

6.5.4 ATOMISTIC MODELS

An explanation for most of the electric double-layer system uses classical and quantum molecular dynamic (MD) simulations and uses parallel high-performance calculations that allow the development of realistic models of such systems. Such a classical model satisfactorily describes the ionic subsystem of the electrolyte. But it failed to provide information regarding the materials of the electrode's electron subsystem. Remaining models are simulated based on Monte Carlo (MC) molecular simulation technique.

An MD simulation is considered as a basic platform for the analysis of molecular interactions and is used widely for the electro osmotic flow modelling and to consider the importance of high charge density in EDLCs. MD has been used to study the influence of temperature on capacitance, and detailed information of the capacitive behaviour of room temperature ionic liquid-based supercapacitors possessing various forms of carbon electrodes are provided. Theoretical and experimental work shows that capacitance has a positive, negative, and even complex relationship with temperature. MD simulation on an Onion Like Carbon (OLC) based supercapacitor shows that capacitance increases with temperature while that of on carbon nanotube-based supercapacitors show the weak dependency of capacitance on temperature in agreement with experimental data. Similarly, MD simulations by Vatamanu et al. on graphitic-based supercapacitor show a negative dependency of capacitance on temperature [37].

The MC study showed that a change in the counter-ion's position and packing of ions on the charged surface has a significant role in the temperature dependence of the EDLC near a porous charged electrode [38]. MC simulations by Boda and

Henderson, show the dependence of capacitance on temperature as bell-shaped. Future work in molecular modelling should consider ionic transport and the kinetics of different charging/discharging rates, and improvement in the performance of electrodes in order to optimize power and energy densities.

6.5.5 Quantum Models

The Car–Pirandello approach based on ab initio molecular dynamics and electron density functional theory (DFT) is the suitable method of quantum modelling for a consistent exploration of the properties of an electrolyte's ionic subsystem and an electrode's electron-hole subsystem in a single calculation [39]. The main postulate of DFT is that the energy of the electron subsystem is a function of the electron density, and the ground state energy is considered as a minimum. Lankin et al. constructed such a model in which quantum molecular dynamics based on DFT were used in the last stage of calculations to define the exact distribution of charges under applied voltage. It described the electrical double layer (EDL) at the electrode (carbon)-electrolyte interface and also the properties of the electrolyte and electron hole subsystem of the solid electrode [40].

Based on DFT, several studies have been conducted like the joint DFT for electrodes used for pseudocapacitors and hybrid capacitors. Recently, DFT was employed to determine the quantum capacitance (QC) of different electrodes using the density of states for supercapacitor applications. The specific capacitance of an electrode can be calculated by the following equation (6.18) and represents the importance of QC:

$$\frac{1}{C} = 1/C_{\mathrm{EDL}} + 1/QC \qquad (6.18)$$

where C_{EDL} is the EDL capacitance.

It is clear from the equation that specific capacitance will be increased on increasing QC, thus the use of electrode materials possessing high QC will significantly increase their capacitance. Classical density functional theory (CDFT) is widely employed to study the structure of EDL in an aqueous electrolyte but not used for the study of differential capacitance of non-aqueous electrolyte/electrode interface and the dependency of pore size on EDLC. CDFT approaches are widely used for the complete energy storage process in EDLC, i.e., from pore characterization to structures of EDL, and shows a good agreement with the experimental observation, hence verifying its importance in the ionic system [25].

6.5.6 Simplified Analytical Models

Simplified analytical models use a set of mathematical equations like partial differential equations for the characterization of electrical performance that defines the physical properties like the flow of charged particles and the reaction rate at the interface of a supercapacitor [20]. This approach consists of only low experimentation and is highly flexible so that new sets of equations and parameters can be included, and it could not predict the ageing process as the analytical models do not consider the non-homogeneous character of electrolytes at the interface and thermal-coupling variables. The most commonly used analytical models are RC circuits, the multi-branch model, and the dynamic model [41].

Kazaryan et al. constructed a simplified analytical model for asymmetric supercapacitors for the calculations as well as for the improvement and control of the energy, capacitance, conductivity, power, etc., for the secure and lengthier operation of different design and types of heterogeneous EDL electrochemical supercapacitors, on accounting for the electrical, physical, and electrochemical properties of materials, designs, and spatial structures of electrodes and separator simultaneously [42]. An electric circuit model has to be simplified in order to nullify the complexity in mathematical representation as well as for an accurate prediction of supercapacitor behaviour.

6.6   SUMMARY

Globally, countless inconsistencies persist in the evaluation of the electrochemical performances of supercapacitors due to the differences in testing and evaluation approaches, difference in testing instruments, electrode-fabrication, experimental set-up differences like three-electrode and two-electrode configurations, and the various test conditions applied. Some standard international protocols must be utilized for the performance evaluation of energy storage devices such as supercapacitors and batteries. One of the major challenges in the field of energy storage is to develop supercapacitors with higher energy density and batteries with higher power density. But it's necessary to establish appropriate methods and techniques for the adequate evaluation of the electrochemical performances of electrode materials in the field of energy storage.

REFERENCES

1. Chen, G. Z. (2017). Supercapacitor and Supercapattery as Emerging Electrochemical Energy Stores. *International Materials Reviews 62*(4), 173–202.
2. Conway, B. B. (1997). The Role and Utilization of Pseudocapacitance for Energy Storage by Supercapacitors, *Power Sources 66*, 1–14.
3. Ratha, S., & Samantara, A. K. (2018). Characterization and Performance Evaluation of Supercapacitor. In *Supercapacitor: Instrumentation, Measurement and Performance Evaluation Techniques* (pp. 23–43). Springer, Singapore.
4. Alano, J. H., da Silva Rodrigues, L., Maron, G. K., da Silveira Noremberg, B., de Oliveira, A. D., Paniz, O., ... & Carreno, N. L. V. (2019). A Simple and Complete Supercapacitor Characterization System Using a Programmable Sourcemeter. *Orbital: The Electronic Journal of Chemistry 11*(2), 133–141.
5. Kampouris, D. K., Ji, X., Randviir, E. P., & Banks, C. E. (2015). A New Approach for the Improved Interpretation of Capacitance Measurements for Materials Utilised in Energy Storage. *Rsc Advances 5*(17), 12782–12791.
6. Negroiu, R., P. Svasta, C. Pirvu, Al. Vasile, and C. Marghescu. (2017). Electrochemical Impedance Spectroscopy for Different Types of Supercapacitors. In 2017 40th International Spring Seminar on Electronics Technology (ISSE), 1–4.
7. Scibioh, M. Aulice, and B. Viswanathan. (2020). Characterization Methods for Supercapacitors. *Materials for Supercapacitor Applications* 315–72.

8. Choudhary, Y. S., Jothi, L., & Nageswaran, G. (2017). Electrochemical Characterization. In *Spectroscopic Methods for Nanomaterials Characterization* (pp. 19–54). Elsevier.

9. Choi, Hyun-Jung, Sun-Min Jung, Jeong-Min Seo, Dong Wook Chang, Liming Dai, & Jong-Beom Baek. (2012) Graphene for Energy Conversion and Storage in Fuel Cells and Supercapacitors. *Nano Energy 1*, 534–51.

10. Banerjee, A. N., Anitha, V. C., & Joo, S. W. (2017). Improved Electrochemical Properties of Morphology-Controlled Titania/Titanate Nanostructures Prepared by in-situ Hydrothermal Surface Modification of Self-source Ti Substrate for High-Performance Supercapacitors. *Scientific Reports 7*(1), 1–20.

11. Liu Tianyu, *Exploration Of Carbonaceous Materials For Supercapacitors*, University Of California Santa Cruz, June 2017

12. Zhang, Z. J., Han, B., Zhao, K. Y., Gao, M. H., Wang, Z. Q., Yang, X. M., & Chen, X. Y. (2020). Surface Modification of Carbon Materials by Nitrogen/Phosphorus Co-doping as well as Redox Additive of Ferrous Ion for Cooperatively Boosting the Performance of Supercapacitors. *Ionics 26*(6), 3027–3039.

13. Alano, J. H., da Silva Rodrigues, L., Maron, G. K., da Silveira Noremberg, B., de Oliveira, A. D., Paniz, O., ... & Carreno, N. L. V. (2019). A Simple and Complete Supercapacitor Characterization System Using a Programmable Sourcemeter. *Orbital: The Electronic Journal of Chemistry 11*(2), 133–141.

14. Cole, K. S., & Robert H. C. (1941). Dispersion and Absorption in Dielectrics I. Alternating Current Characteristics. *The Journal of Chemical Physics 9*, 341–351.

15. Gabrielli Claude, G., *Identification of Electrochemical Processes by Frequency Response Analysis Technical Report Number 004/83*, Universite P. et M. Curie, France, 3rd March 1998.

16. Lu, M. (2013). *Supercapacitors: Materials, Systems, and Applications*. John Wiley & Sons.

17. Halper, M. S., & Ellenbogen, J. C. (2006). Supercapacitors: A Brief Overview. In The MITRE Corporation, McLean, Virginia, USA, *1*.

18. Masarapu, C., Zeng, H. F., Hung, K. H., & Wei, B. (2009). Effect of Temperature on the Capacitance of Carbon Nanotube Supercapacitors. *ACS Nano 3*(8), 2199–2206.

19. Mei, B. A., Munteshari, O., Lau, J., Dunn, B., & Pilon, L. (2018). Physical Interpretations of Nyquist Plots for EDLC Electrodes and Devices. *The Journal of Physical Chemistry C 122*(1), 194–206.

20. Berrueta, A., Alfredo U., Idoia San M., Ali E., & Pablo S. (2019). Supercapacitors: Electrical Characteristics, Modeling, Applications, and Future Trends. *IEEE Access 7*, 50869–96.

21. Zhang, L., Hu, X., Wang, Z., Sun, F., & Dorrell, D. G. (2018). A Review of Supercapacitor Modeling, Estimation, and Applications: A Control/Management Perspective. *Renewable and Sustainable Energy Reviews 81*, 1868–1878.

22. Castiglia, V., Campagna, N., Spataro, C., Nevoloso, C., Viola, F., & Miceli, R. (2020). Modelling, Simulation and Characterization of a Supercapacitor. In 2020 IEEE 20th Mediterranean Electrotechnical Conference (MELECON) (pp. 46–51). IEEE.

23. Castiglia, V., N. Campagna, A. O. Di Tommaso, R. Miceli, F. Pellitteri, C. Puccio, & F. Viola. (2020). Modelling, Simulation and Characterization of a Supercapacitor in Automotive Applications." In 2020 Fifteenth International Conference on Ecological Vehicles and Renewable Energies (EVER), 1–6. Monte-Carlo, Monaco: IEEE.

24. Devillers, N., Samir, J., Marie-Cécile, P., Daniel B., & Frédéric, G. (2014). Review of Characterization Methods for Supercapacitor Modelling. *Journal of Power Sources 246*, 596–608.

25. Bharti, A. K., Gulzar, A., Meenal, G., Patrizia, B., Ravikant, A., Ramesh, C., & Yogesh, K. (2021). Theories and Models of Supercapacitors with Recent Advancements: Impact and Interpretations. *Nano Express 2*, 022004.

26. Kroupa, M., Gregory, J. O., & Juraj, K. (2016). Modelling of Supercapacitors: Factors Influencing Performance." *Journal of The Electrochemical Society 163*, A2475–87.

27. Entremont, A., & Laurent P. (2014). First-Principles Thermal Modeling of Electric Double Layer Capacitors under Constant-Current Cycling. *Journal of Power Sources 246*, 887–98.

28. Zubieta, L., & R. Bonert. (2000). Characterization of Double-Layer Capacitors for Power Electronics Applications. *IEEE Transactions on Industry Applications 36*, 199–205.

29. Diab, Y., P. Venet, H. G., & Rojat, G. (2009). Self-Discharge Characterization and Modeling of Electrochemical Capacitor Used for Power Electronics Applications. *IEEE Transactions on Power Electronics 24*, 510–17.

30. Ghosh, A., Viet T. L., Jung J. B., & Young H. L (2013). TLM-PSD Model for Optimization of Energy and Power Density of Vertically Aligned Carbon Nanotube Supercapacitor. *Scientific Reports 3*, 2939.

31. Gonçalves, R., W. A. Christinelli, A. B. Trench, A. Cuesta, & E. C. Pereira. (2017). Properties Improvement of Poly(o-Methoxyaniline) Based Supercapacitors: Experimental and Theoretical Behaviour Study of Self-Doping Effect. *Electrochimica Acta 228* (February): 57–65.

32. Levie, R. de. (1963). On Porous Electrodes in Electrolyte Solutions. *Electrochimica Acta 8*, no. 10, 751–80.

33. Kurzweil, P., & H. J. Fischle. (2004). A New Monitoring Method for Electrochemical Aggregates by Impedance Spectroscopy." *Journal of Power Sources 127*, 331–40.

34. Pilon, L., Hainan, W., & Anna d', E. (2015). Recent Advances in Continuum Modeling of Interfacial and Transport Phenomena in Electric Double Layer Capacitors. *Journal of The Electrochemical Society 162*, A5158–78.

35. Wang, H., & Laurent P. (2013). Mesoscale Modeling of Electric Double Layer Capacitors with Three-Dimensional Ordered Structures." *Journal of Power Sources 221*, 252–60.

36. Woo, S.-W., Kaoru D., Hiroyuki N., & Kiyoshi K. (2008). Preparation of Three Dimensionally Ordered Macroporous Carbon with Mesoporous Walls for Electric Double-Layer Capacitors. *Journal of Materials Chemistry 18*, 1674.

37. Vatamanu, J., Oleg B., & Grant D. S. (2011). Molecular Simulations of the Electric Double Layer Structure, Differential Capacitance, and Charging Kinetics for N-Methyl- N -Propylpyrrolidinium Bis(Fluorosulfonyl)Imide at Graphite Electrodes. *The Journal of Physical Chemistry B 115*, 3073–84.

38. Reszko-Zygmunt, J., S. Sokołowski, D. Henderson, & D. Boda. (2005). Temperature Dependence of the Double Layer Capacitance for the Restricted Primitive Model of an Electrolyte Solution from a Density Functional Approach. *The Journal of Chemical Physics 122*, 084504.

39. Pillay, B., & John N. (1996). The Influence of Side Reactions on the Performance of Electrochemical Double-Layer Capacitors. *Journal of the Electrochemical Society 143*, 1806–14.

40. Lankin, A. V., G. E. Norman, & V. V. Stegailov. (2010). Atomistic Simulation of the Interaction of an Electrolyte with Graphite Nanostructures in Perspective Supercapacitors. *High Temperature 48*, 837–45.

41. Shi, L., & M. L. Crow. (2008). Comparison of Ultracapacitor Electric Circuit Models. In *2008 IEEE Power and Energy Society General Meeting - Conversion and Delivery of Electrical Energy in the 21st Century*, 1–6. Pittsburgh, PA: IEEE.

42. Kazaryan, S. A., S. N. Razumov, S. V. Litvinenko, G. G. Kharisov, & V. I. Kogan. (2006). Mathematical Model of Heterogeneous Electrochemical Capacitors and Calculation of Their Parameters. *Journal of the Electrochemical Society 153*, A1655.

7 Design, Fabrication, and Operation

Sudhakar Y. N., Shilpa M. P., Shivakumar Shetty,
Sreejesh M., Shridhar Mundinamani,
Margandan Bhagiyalakshmi, and Gurumurthy S. C.

CONTENTS

7.1 INTRODUCTION

The concept of supercapacitors is blooming as a complementary technology or replacement of batteries in sectors that demand high power density, high cycle life, fast charging, grid-scale power stability, and enhanced power-to-weight ratio, and the high industrial demand makes it highly significant to look into the design and operation of fully functional supercapacitor prototypes. The design and assembly of a supercapacitor are planned based on the appropriate rating and application. A complete cell consists of a pair of electrodes insulated by a separator, sandwiched under constant pressure between current collector electrodes. The most common commercialized supercapacitor is in the shape of a cylinder or of a coin. The fundamentals of all supercapacitors are consistent, while their packaging differs based on their manufacturers and applications.

7.2 CONSIDERATIONS AND TRENDS FOR SINGLE-CELL SUPERCAPACITORS

The simplest way to construct a single cell is by using a coin cell. Cylindrical and prismatic design are other widely followed ways of fabrication of supercapacitors.

DOI: 10.1201/9781003258384-7

One of the critical factors in making a single cell is the choice of suitable electrode materials, which should have a higher surface area and provide higher electrochemical performance. To improve the energy density of the supercapacitors, it is vital to increase the surface area of the electrodes. Furthermore, researchers need to identify the new electrode active material and electrolyte materials which can give energy density as that of the battery. The voltage and net capacitance can be varied by connecting the supercapacitor cells either in series, parallel, or both.

Due to the rigid electrodes, the traditional supercapacitors are limited to the shape of the device. With the rapid rise of wearable and portable electronic devices, the trend is moving towards flexible energy storage devices, which has attracted the attention of researchers. So, a flexible supercapacitor is one of the next-generation portable energy storage devices. To achieve it is essential to the design and manufacture of flexible electrodes, fluid collectors, and packaging shells which give the supercapacitor an option to fit into any design of shape and size, thus improving their potential for applications in flexible and wearable fields [1].

Considering the fact that the single supercapacitor cell cannot meet the requirements as those of a single battery at present, improvising the materials and design is required [2]. Currently, miniaturized electronic devices are the focus of interest amongst the consumers, hence micro energy devices are in great demand [3]. Advancement in microsupercapacitors (MSCs) will endow applications in nanorobotics, wearable devices, portable sensors, etc. [4]. Stamping MSCs with planar interdigital configurations is also promising printing design [5]. Another design technique is a 3D printing which involves fused deposition modelling (FDM) to prepare the frame for a supercapacitor [6]. So, at present growing demand and user-friendly design for requirements of the supercapacitor, data-driven software modelling is emerging. The Gaussian process regression and molecular dynamics (MD) simulations envisage and optimize the design and efficiency of the electrodes [7].

7.2.1 Coin Cell Supercapacitor

Coin cell supercapacitors are very compact and possess high capacitance. It is associated with electric double-layer storage that can store energy more and longer than conventional batteries. A coin cell is known to withstand nearly 500,000 charge/discharge cycles or stay up to 10 years without a compromise in energy in each cycle. Based on the termination requirements they are of three types, radial, horizontal, and vertical configurations. Usually, a single coin cell is designed containing carbon material as anode and cathode material and organic electrolyte. This cell can be stacked in series to achieve 5.5–11 V using conductive rubber and sealant to meet the requirements. The surge voltage is usually 5.8 V and capacity range of 0.1–1.5 F. The operating temperature range can be from −25 to −70 °C. Commercial coin cell supercapacitors have capacity attenuation of ≤30% and equivalent series resistance (ESR) change ≤4 times. Although coin cell batteries are well known and used widely, coin cell supercapacitors have different fabrication and testing requirements. Mainly, the rate of charge/discharge is faster in supercapacitors than batteries. During fabrication the amount of pressure applied to stack the materials, inclusion of conductive fillers, thickness of current collectors, and package is to be considered for enhancing the rate capability of a supercapacitor [8] [9]. Coin cell supercapacitors are used in routers and switches, utility meters, computers, and peripherals.

7.2.2 CYLINDRICAL CELL SUPERCAPACITOR

A cylindrical cell supercapacitor features a cylindrical design to wind the anode, cathode, and electrolyte materials inside a cylindrical case [10]. This cell comes initially with 2.7 V and goes out to 3.2 V. The best part is that it can be combined with a battery to produce 4 V with good pulse capability. The capacitance range is from 30 to 220 F with 4 V surge voltage and can operate from −25 to 70°C. They come in either board mounted or threaded terminals but due to their cylindrical shape the empty space will always be associated in the region where the corners are located in other designs. Hence, the application in compact portable devices is not recommended. Nevertheless, they are used in wireless sensors and actuators, peak power support for drug delivery systems, and a short-term battery replacement.

7.2.3 PRISMATIC CELL SUPERCAPACITOR

A prismatic cell supercapacitor is a "pouch like" design known for very low-profile footprint design which may go up to 0.4 mm thickness [11, 12]. This design can withstand high pulse power and has very low ESR. Moreover, it can operate at a temperature range of −40–70°C when compared to other designs. The energy density and power density are delivered from a wafer-thin, robust, compact, and lightweight design package. The advantage is that it can be either be used singly or in combination with primary/secondary batteries which gives instantaneous backup during battery failure. It has high radiation efficiency and is easy to implement in portable devices. Therefore, it is mainly applied for assisting battery, energy, and power holdup, handling pulse power, and energy storage [13, 14]. Furthermore, incorporation of activated carbon in the positive electrode, instead of metal oxides like lithium cobalt oxide will result in inhibiting thermal runaway reaction.

7.3 PARAMETERS AFFECTING PERFORMANCE

To meet the demand in the fields of varying power input to temperature, the performance of a supercapacitor can be tailored by tuning the time constant, operating voltage, specific capacitance, energy and power density, and ESR (see Figure 7.1).

The three main electrical characterizations for supercapacitors involve electro-chemical impedance spectroscopy (EIS), cyclic voltametric (CV), and galvanostatic charge/discharge (GCD) methods. CV and GCD output from a battery, asymmetric supercapacitor, and a hybrid capacitor. In a battery, the CV and GCD exhibit redox behaviour showing charge/discharge plateaus. The rectangular-shaped CV curve and triangular mirror image-shaped GCD curve result from capacitive behaviour in asymmetric supercapacitors. The hybrid behaviour obtained by the combination

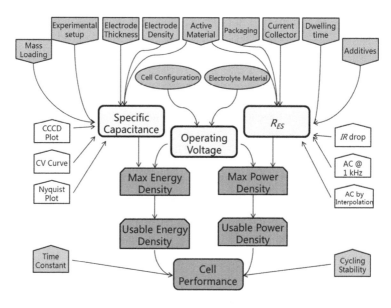

FIGURE 7.1   An illustration of key performance metrics, test methods, and major factors affecting the evaluation of supercapacitors. *Reproduced with permission from [15], Wiley publication. Moreover, for fabrication of products based on commercial requirements material, weight, design, flexibility, operating temperature, operating current, and voltage are to be fixed. The performance criteria of supercapacitors is shown in Figure. 7.1.*

of capacitive and redox properties which deviates from ideal capacitive behaviour is observed in the CV and GCD curves.

Supercapacitors are known for having low ESR or initial leakage current because when it is uncharged a sudden power is drawn from the supply. This instance would be fine in small equipment, but when in grids or sensitive circuits will lead to significant voltage drop due to instantaneous power drawn by supercapacitor, system failure, and software reset. Hence, charge-limiting circuits are necessary to control the fluctuations. Supercapacitor performance depends on factors such as specific capacitance, charge/discharge stability, and time constant. So, the electrode material is designed considering the following aspects. (a) Surface area: the higher the surface area the higher the specific capacitance because it can store more charges. (b) Enhanced electronic and ionic conductivity reduces the loss during scanning GCD and CV at high current density and scan rates, respectively. Binder-free electrode coating on the electrode reduces electronic losses. The increased porous structure will provide channels for ionic conductivity thereby increasing the specific capacitance. (c) Mechanical and chemical stability: the reason for low performance is due to side reactions of active materials, phase change and brittle surfaces. A design comprising from 0D to 3D structures is one of the emerging domains [16]. Designing an electrolyte for its utilization in a supercapacitor is based on its performance expectations. The main indicators are similar to electrode materials which are capacitance, charge storage, rate performance, time constants, voltage window, cyclability, characteristic resistances, and safety. Furthermore, factors affecting the electrode/electrolyte interface region and the bulk of electrolytes such as ionic conductivity and mobility, solvation, diffusion coefficient, the radius of hydrated spheres, dielectric constant, electrochemical stability, thermal stability, and dispersion interaction as to be studied before designing an optimized electrolyte [17]. The electrolyte choice depends on design and application need as there are three types, liquid electrolytes, solid-state or quasi-solid-state electrolyte, and redox-active electrolytes. In liquid electrolytes: water-in-salt electrolyte, aqueous electrolyte (acid, alkaline, neutral salts), and non-aqueous (organic, ionic liquid, mixtures). Solid electrolytes consist of mainly dry solid [18], gel polymer [19], and inorganic materials. Redox electrolytes consist of materials like aqueous KI, ionic liquid ([EMIM][I]$^+$, organic $(Et_4N)_2B_{12}F_{11}H$/propylene carbonate, and gel polymer (PVA + KI in H_2O or rGO/conducting polymer [20, 21].

7.4 OPERATION OF FUNCTIONAL SUPERCAPACITOR

7.4.1 SELF-DISCHARGING

Despite their many benefits, such as fast charging and discharging, long cycle life, environmental friendliness, high power density, and high safety, supercapacitors have some demerits that may limit their practical employment. Self-discharge occurs when the voltage of charged supercapacitors drops spontaneously over time. The phenomenon of self-discharge adversely affects the characteristics of the capacitor i.e., power and energy density. Hence it is essential to comprehend the mechanism of self-discharge to enhance the efficiency of supercapacitors.

In comparison to discharged or partially discharged supercapacitors, fully charged supercapacitors have a higher free energy state. Therefore, the virtual driving force due to certain processes may provide a path for the spontaneous decrease of the energy. However, there is no mechanism for self-discharge for an ideally polarizable electrode, and hence the extent of charge is constant with respect to time. However, such an ideal behaviour is practically impossible in most cases. Generally, there are three phenomena that can control the self-discharge of supercapacitors.

i.   Ohmic leakage: it is one of the explicit phenomena in which a resistive path between the electrodes emerging due to construction faults results in self-discharge. Ohmic leakage resistance and capacitance influence the rate of self-discharge in a supercapacitor. It might be due to incomplete sealing or inter-electrode contacts [22, 23].

ii. Faradic reactions: the presence of certain impurities can cause oxidation or reduction reactions resulting in the discharging of the capacitors. Electrolyte anions balance the positive charge on the positive electrode in the double layer. Any impurity species that oxidize on the positive electrode will pass their electrons across the double layer into the electrode surface, lowering the positive charge on the electrode surface resulting in self-discharge.

Similarly, electrolyte cations balance the negative charge on the negative electrode and therefore any reduction reaction on the surface tends to the self-discharge.

For, example, in sulphuric acid-based electrolyte, iron is the most common contaminant, and it can oxidize on the surface of positive metal electrodes, thus causing self-discharge. Similarly, in the case of carbon-based materials, on the surface of negative electrodes, oxygen-containing functional groups can cause a reduction reaction on the negative electrode resulting in self-discharge.

Chemical reactions for the above example are given below:

$$M^+ + HSO_4^- + Fe^{2+} \rightarrow M + HSO_4^- + Fe^{3+} \tag{7.1}$$

$$4M^- + H^+ + O_2 \rightarrow 2H_2O + 4M \tag{7.2}$$

Based on the concentration of the reactants, there are two types of faradic reactions for supercapacitors: activation-controlled reactions and diffusion-controlled reactions. When the reactant species concentration is high enough, the influence of concentration on the faradic reaction can be omitted. Hence the activation-controlled reaction dominates the process of faradic reaction-induced self-discharge. On other hand, when the concentration is low enough, the time taken by the reactants to transport through the electric double layer takes a long time to get reduced or oxidized on the surface of the electrode. Hence, in this case, a diffusion-controlled reaction dominates the self-discharge mechanism [23–25].

iii. Redistribution of charges: after the removal of the external power supply to the electrode, charges will not have sufficient time to distribute evenly in the pores of the porous electrode materials like carbon-based materials, and hence, most of them will adhere to the tip of the pores and later they proceed towards the depth of the pores on account of concentration gradient of charges. As a result, the tip of the pore will experience a potential drop causing self-discharge of the electrodes. Here, the size of the pore plays a crucial role [25].

iv. Hybrid mechanism: the complex nature of each component of supercapacitors causes the self-discharge process to be regulated by the combination of different mechanisms like leakage current, faradic reactions, and charge redistribution [25].

The self-discharge potential is assessed by a two- and three-electrode system with time in an open circuit [22].

Various strategies are followed to suppress the self-discharging of the supercapacitors so that the efficiency of the supercapacitors can be enhanced as there is a possibility of 5–60% of self- discharge within two weeks. Some of the reported methods to reduce the self-discharge rate are modifying the electrode material, modulating the electrolyte, and tuning the separator.

Carbon allotropes like graphene derivatives, carbon nanotubes, activated carbon, etc., are the most widely used electrode materials because of their high surface area and porosity for the fast movement of ions. However, the presence of impurities and functionalities can cause self-discharge by ohmic leakage and faradic reactions. In addition, porous structure of some of the carbon-based material causes charge redistribution leading to the self-discharging of electrode. Hence there is a need to overcome these hindrances by modifying the structures. As reported earlier, a number of oxygen functional groups can be reduced by high temperature H_2 treatment as well as treating with molten salt at 600°C and hence it can suppress the self-discharge. Besides, hybridizing with polymeric materials is another approach which can reduce the self-discharge rate controlled by ohmic leakage and a parasitic faradic reaction [25].

The rate of self-discharge depends on the electrolyte as well. The size of the ions and viscosity of the electrolyte has a large impact on charge redistribution and faradic reactions. Aqueous electrolyte is the most used low energy density due to a narrow window limiting its practical utilization. The decreased amount of dissolved oxygen in the electrolyte and contamination adversely affects the supercapacitor by accelerating the self-discharge. A reduction in oxygen concentration in the electrolyte can be resolved by treating with sulphuric acid in the presence of O_2 and H_2 gas. Iron is the most common contaminant found in electrolyte and it can cause self-discharge by faradic reactions. Combination in the aqueous solvent of certain additives such as clay and potassium acetate is another way to hamper the self-discharge. Recently organic electrolytes, i.e., combination of an organic solvent with a salt are favoured in commercial applications due to their high and stable voltage window. A relatively large size organic molecule makes it difficult to approach the electrode surface. The high viscosity of these organic solvents makes the supercapacitor more vulnerable to temperature. The addition of certain materials that can decrease the viscosity of the solvent can decrease the rate of self-discharge.

The contribution of separators towards the self-discharge is significant and cannot be neglected. A Nafion separator is quite commonly used due to its long life, electrochemical stability, and selectivity in allowing migration of H_2 of ions only. Even if the shuttle effect is inhibited to a certain extent in the latter separator, it is unsatisfactory. Modifying the separator and thus inducing either positive or negative charges is one of the techniques to inhibit the corresponding shuttle effect [25].

7.4.2 CELL AGEING AND VOLTAGE DECAY

Supercapacitors undergo ageing after a prolonged period resulting in slower charging and discharging time, decreased capacitance, and increased resistance. As per industry standards, the life of supercapacitors is specified by 30% decrement in capacitance and 100% increment in ESR. The performance of supercapacitors degrades over a period and after a certain time its performance does not suit the application requirements. The deterioration in performance can also be indicated by a decreasing of the energy, an increase in power loses, and some of the macroscopic phenomena such as gas evolution, electrode swelling, and loss of elements due to redox reactions.

Voltage decay is a result of cell ageing. Charging voltage and temperature influence the cell ageing and voltage decay significantly and the voltage decay due to self-discharge is not treated as voltage decay. A slight increase above the rated voltage and temperature can increase the rate of electrochemical reactions, accountable for cell ageing. An increase in the rate of adverse chemical activities such as electrolyte and electrode decomposition is enhanced at elevated temperature and voltage. In carbonaceous electrodes, containing metal and oxygen functionalities, decomposition products obtained from redox decomposition of electrolytes can block the pores and reduce the capacitance. Moreover, the pores of the separator get blocked due to the decomposition products thus causing increased resistance. Water contamination in organic electrolyte is another factor that can cause macroscopic affects [26, 27].

7.5 SUPERCAPATTERY

Low power density in batteries and low energy density in supercapacitors has led to the new area of research which offers the hybridization of batteries and supercapacitors into a single device. A supercapattery is a generic term for various hybrid devices combining the merits of a rechargeable battery and supercapacitor as it is considered a third-generation energy storage device. Their performance is comparable to that of a supercapacitor with a higher energy capacity and a rechargeable battery, but with more power extended charge/discharge durability and capabilities. A supercapattery is the combination of a Nernstian negatrode and a capacitive positrode, i.e., it is accomplished by pairing a supercapacitor electrode with a battery electrode in a balanced manner, making the electrodes from active materials capable of both capacitive and Nernstian charge storage or adding redox species to the supercapacitor electrolyte so that the device can store charge using both capacitive and Nernstian mechanisms. In the mid-1990s Varakin et al. [28] reported a single device which had a battery electrode (NiO_2) and capacitor electrode (carbon fibre) that showed eight to ten times more performance than a symmetric device. The key to enhanced cycle life and potential is to combine suitable battery and capacitor hybrid electrodes [29].

In supercapattery-type electrode systems, the rate capability is less than pure redox electrode material as it involves faradic reactions occurring deep inside the crystal structure. Table 7.1 depicts the capattery device with different electrode materials and electrolytes.

In the case of scalable supercapatteries, electrode materials and fabrication methods are equally important in the performance of the resulting devices. For instance, bipolar stacking with multiple supercapattery cells can achieve high energy density storage by reducing almost half of the auxiliary materials (current collectors) by ensuring the permeability of the bipolar plates. Ultimately, supercapatteries have attracted much attention as they can balance one device's energy capacity and power capability. The advancement of battery and supercapacitor materials can aid the progress of supercapatteries in this regard.

TABLE 7.1

Summary of Pairing the Electrode Materials of Different Charge Storage Mechanisms* into Supercapacitor, Supercapattery, Supercabattery, and Battery, and the Performance Metrics of the Representative Cells Using Different Electrolytes [32]

Negatrode	Electrolyte	Positrode	References
Li	IL + LiClO$_4$	Act-C (activated carbon)	[30]
Li	PEO-LiTFSI I LTAP I 1.0 M LiCl aq.	MnO$_2$	[30]
C-MnO$_y$-CNT	LiPF$_6$	CNT@MnO$_x$	[30]
Activated carbon	6.0 M aqueous KOH	Ni(OH)$_2$/CuCo$_2$S$_4$/Ni	[31]
Act-C	KCl	PPy-CNT	[32]
G/AC	KOH	MXene/NiS	[33]

7.6 STACK MANUFACTURING AND CONSTRUCTION

Supercapacitors are used in applications where there are often high load-current fluctuations and high voltage demand. Stack manufacturing of supercapacitors either in series or parallel or a combination of both are necessary to meet this intended application [34, 35]. Accordingly, the manufacturers have designed the stacking of supercapacitors as coin, pouch, or cylindrical forms. The materials and shape used during the manufacturing such as sealants, casings, and electrode materials must meet temperatures, operating voltages, and power requirements and the stacking method must be well defined. The major challenge is to reduce the weight-to-energy ratio and improve energy and power densities. Usually, series stacking is preferred over parallel stacking as it occupies less space and reduces the weight-to-energy ratio. The maximum stored energy (W_{max}) in the stacked supercapacitor which gives information about maximum energy storage during charging and energy deliverable to the load is shown in Equation (7.3):

$$W_{max} = \frac{C_{eq}U_{max}}{2} \tag{7.3}$$

where W_{max} is the maximum energy storage capacity, C_{eq} is equivalent capacitance of the supercapacitor stack, U_{max} is maximum voltage of the supercapacitor stack [36].

Some of the stacking designs, like the bipolar design, minimize the volume of the stack, resistance from solder joints, and interconnects, thereby enhancing the supercapacitor performance [37]. In bipolar design a coat with electroactive material on both sides of a highly conducting plate are stacked using a separator. One side of a coated conducting plate acts as the anode while other side as the cathode. Multiple coated conducting plates are sandwiched a separator with gasket and sealant to avoid a short circuit. The space between the electrodes is filled with either liquid electrolyte or solid polymer electrolyte. A bipolar supercapacitor fabricated using graphite/vertically aligned carbon nanotubes (VACNTs) electrodes

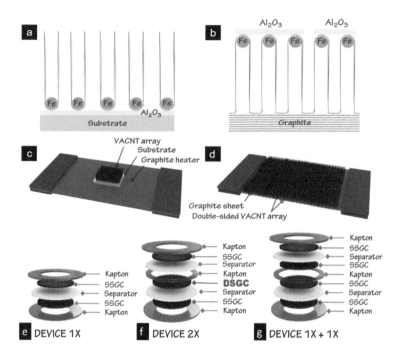

FIGURE 7.2 (a) Schematic of VACNTs grown using conventional CVD, with Fe catalyst and Al_2O_3 diffusion barrier layer beneath. (b) VACNTs grown using odako-growth, with Fe and Al_2O_3 on top of the VACNT array. (c) Conventional cold-wall VACNTs CVD setup, with VACNTs grown on a Si chip placed on a graphite heater. (d) Joule heating CVD setup, with VACNTs grown directly on the graphite foil used as the heater. Finally, schematics of devices (e) 1×, (f) 2× and, (g) 1× + 1×. *Reproduced with permission from [38], Elsevier publication.*

by a chemical vapour deposition (CVD) setup is shown in Figure 7.2. When more working voltage is required, the supercapacitors are connected in series; however, this form of stacking decreases the capacitances and increases the ESR – moreover, it is essential to ensure the equal distribution of cell voltage. The cells are connected in parallel for higher energy density or power density, increasing the capacitance and decreasing ESR. Series stacking is more feasible than parallel stacking during stack creation and packing. Individually packed cells are serially incorporated into a multi-unit stack by external metal bars or soldering connecting the cells to a printed circuit board.

Uncased jelly rolls may be used to save space and weight in the package. Stacking can result in large-scale resistance due to electrode resistance and inter-cell contact when scaling the module designs. It may also lead to uneven charge distribution, which can be rectified by passive and active balancing. [39]. The voltage provided to each cell must be limited to its rated voltage, which necessitates a cell voltage balancing circuit. Controlling overvoltage between supercapacitor cells and balancing the charge on armatures may be done in various ways. The following balancing and overvoltage protection circuit types are used in practical applications: resistive dividers, dividers with switched resistors, Zener diode peak limiter, rectifier diode peak limiter, electronic balancing circuits [40].

The simplest circuit for voltage balancing is a resistive divider composed of parallelly connected identical resistors ensuring proper voltage distribution after charging. The downside of this method is that when the charging circuit is turned off, the capacitor discharges through the balancing resistor. By removing balancing resistors after the capacitor has charged, dividers with switched resistors overcome the disadvantage mentioned above. When the diode's threshold voltage is achieved, parallel-linked Zener diodes are used to restrict voltage. When the voltage drop across cells is less than the diode threshold voltage, the current value of this circuit is insignificant. In forward conduction, rectifier diodes (p-n junctions) can also be used to create voltage balance. This method can be effectively restricted when the current travels through a linked capacitor and reaches a set potential. The energy efficiency of this technology is as high as 92%. Electronic balancing circuits incorporating DC/DC converters that assure consistent voltage distribution between the cells are linked between neighbouring supercapacitors. These converters actively equalize the voltages of the cells. In this situation, power losses are decreased (the converter's power losses are eliminated), but expenses increase (hardware implementation and control are costly). One could summarize that a supercapacitive balancing solution is different for dissipative balancing circuits as well as non-dissipative solutions. The dissipation balancing circuit is divided into passive balancing and active balancing. Passive balancing makes use of parallel resistor and Zener diodes, while active balancing utilizes switched parallel resistors as well as op-amp output current. In non-dissipative balancing circuits, there is indirect balancing that uses a buck-boost converter/charge pump, whereas direct balancing comprises a flyback converter and a forward converter. The thumb rule for EDLC supercapacitors derived from a manufacturer datasheet states that the cell lifetime doubles in a specified voltage range with every 0.2 V decrease [41].

The speculated voltage is halved as voltage increases every 0.1-fold balancing significantly affects the cell lifetime and efficiency. The lifetime prediction of cell with passive and active balancing is 50–80% and 90–95%, respectively. The efficiency for passive balancing is 70–80% and 90–95% for active balancing. The lifetime prediction and efficiency of direct/indirect balancing are 95–100% and 90–95%, respectively. The applied voltage by the number of connected cells should not surpass 85% of the rated cell voltage without external voltage dividers. Aqueous electrolyte-based low voltage (approx. 1.8 V) supercapacitors do not require balancing on higher voltage modules as the change in applied voltage on a large number of interconnected cells is not crucial [42].

7.7   SUMMARY AND OUTLOOK

The fundamental aspects of supercapacitor fabrication and properties like single-cell manufacturing, self-discharging, cell ageing, voltage decay, stack manufacturing, etc., are discussed in this chapter. The advancement of single-cell manufacturing via micro-printing, 3D printing, computer modelling, and asymmetric combination of electrodes has attracted the attention of lead researchers. Nevertheless, due to the high voltage operations and fluctuations in the current, supercapacitors are subjected to self-discharging, ageing, and voltage decay phenomenon which need to be addressed during fabrication. A supercapattery device, as a hybrid device, is of great interest and serves the need of high power and energy density when compared to batteries and supercapacitors. Nonetheless, the scope for commercialization of this device is promising. The requirement of constant high voltage without affecting the grid for long charge/discharge cycles, stack manufacturing is increasing instead of conventional solder joints either in series or parallel connections. Hence, the performance of supercapacitors basically depends on their design construction materials and working conditions.

REFERENCES

1. Shao, Y., El-Kady, M. F., Sun, J., Li, Y., Zhang, Q., Zhu, M, & Kaner, R. B. (2018). Design and mechanisms of asymmetric supercapacitors. *Chemical Reviews*, 118(18), 9233–9280.
2. Muzaffar, A., Ahamed, M. B., Deshmukh, K., & Thirumalai, J. (2019). A review on recent advances in hybrid supercapacitors: Design, fabrication and applications. *Renewable and Sustainable Energy Reviews*, 101, 123–145.
3. Wang, J., Li, F., Zhu, F., & Schmidt, O. G. (2019). Recent progress in micro-supercapacitor design, integration, and functionalization. *Small Methods*, 3(8), 1800367.
4. Nikonovas, T., Spessa, A., Doerr, S. H., Clay, G. D., & Mezbahuddin, S. (2020). Near-complete loss of fire-resistant primary tropical forest cover in Sumatra and Kalimantan. *Communications Earth & Environment*, 1(1), 1–8.
5. Li, F., Li, Y., Qu, J., Wang, J., Bandari, V. K., Zhu, F., & Schmidt, O. G. (2021). Recent developments of stamped planar micro-supercapacitors: Materials, fabrication and perspectives. *Nano Materials Science*, 3(2), 154–169.
6. Tanwilaisiri, A., Xu, Y., Zhang, R., Harrison, D., Fyson, J., & Areir, M. (2018). Design and fabrication of modular supercapacitors using 3D printing. *Journal of Energy Storage*, 16, 1–7.
7. Patel, A. G., Johnson, L., Arroyave, R., & Lutkenhaus, J. L. (2019). Design of multifunctional supercapacitor electrodes using an informatics approach. *Molecular Systems Design & Engineering*, 4(3), 654–663.
8. Stoller, M. D., Stoller, S. A., Quarles, N., Suk, J. W., Murali, S., Zhu, Y., ... & Ruoff, R. S. (2011). Using coin cells for ultracapacitor electrode material testing. *Journal of Applied Electrochemistry*, 41(6), 681.
9. Miao, J., Zhou, C., Yan, X., Jiang, H., You, M., Zhu, Y., ... & Cheng, X. (2021). Electrochemical performance of an asymmetric coin cell supercapacitor based on marshmallow-like MnO2/carbon cloth in neutral and alkaline electrolytes. *Energy & Fuels*, 35(3), 2766–2774.
10. Lee, J. H., Kim, H. K., Baek, E., Pecht, M., Lee, S. H., & Lee, Y. H. (2016). Improved performance of cylindrical hybrid supercapacitor using activated carbon/niobium doped hydrogen titanate. *Journal of Power Sources*, 301, 348–354.
11. Bhattacharjya, D., Carriazo, D., Ajuria, J., & Villaverde, A. (2019). Study of electrode processing and cell assembly for the optimized performance of supercapacitor in pouch cell configuration. *Journal of Power Sources*, 439, 227106.
12. Poochai, C., Sriprachuabwong, C., Srisamrarn, N., Chuminjak, Y., Lomas, T., Wisitsoraat, A., & Tuantranont, A. (2019). High performance coin-cell and pouch-cell

supercapacitors based on nitrogen-doped reduced graphene oxide electrodes with phenylenediamine-mediated organic electrolyte. *Applied Surface Science*, 489, 989–1001.

13. Lei, C., Fields, R., Wilson, P., Lekakou, C., Amini, N., Tennison, S., ... & Martorana, B. (2021). Development and evaluation of a composite supercapacitor-based 12 V transient start–stop power system for vehicles: Modelling, design and fabrication scaling up. *Proceedings of the Institution of Mechanical Engineers, Part A: Journal of Power and Energy*, 235(4), 914–927.

14. Patil, A. M., An, X., Li, S., Yue, X., Du, X., Yoshida, A., ... & Guan, G. (2021). Fabrication of three-dimensionally heterostructured rGO/WO3· 0.5 H2O@ Cu2S electrodes for high-energy solid-state pouch-type asymmetric supercapacitor. *Chemical Engineering Journal*, 403, 126411.

15. Zhang, S., & Pan, N. (2015). Supercapacitors performance evaluation. *Advanced Energy Materials*, 5(6), 1401401.

16. Yu, Z., Tetard, L., Zhai, L., & Thomas, J. (2015). Supercapacitor electrode materials: Nanostructures from 0 to 3 dimensions. *Energy & Environmental Science*, 8(3), 702–730.

17. Pal, B., Yang, S., Ramesh, S., Thangadurai, V., & Jose, R. (2019). Electrolyte selection for supercapacitive devices: A critical review. *Nanoscale Advances*, 1(10), 3807–3835.

18. Sudhakar, Y. N., Selvakumar, M., Bhat, D. K., & Kumar, S. S. (2014). Reduced graphene oxide derived from used cell graphite and its green fabrication as an eco-friendly supercapacitor. *RSC Advances*, 4(104), 60039–60051.

19. Sumana, V. S., Sudhakar, Y. N., Anitha, V., & Nagaraja, G. K. (2020). Microcannular electrode/polymer electrolyte interface for high performance supercapacitor. *Electrochimica Acta*, 353, 136558.

20. Sudhakar, Y. N., Smitha, V., Poornesh, P., Ashok, R., & Selvakumar, M. (2015). Conversion of pencil graphite to graphene/polypyrrole nanofiber composite electrodes and its doping effect on the supercapacitive properties. *Polymer Engineering & Science*, 55(9), 2118–2126.

21. Sudhakar, Y. N., Selvakumar, M., & Bhat, D. K. (2018). *Biopolymer Electrolytes: Fundamentals and Applications in Energy Storage*. Elsevier.

22. Conway, B. E. (2013). *Electrochemical Supercapacitors: Scientific Fundamentals and Technological Applications*. Springer Science & Business Media.

23. Conway, B. E., Pell, W. G., & Liu, T. C. (1997). Diagnostic analyses for mechanisms of self-discharge of electrochemical capacitors and batteries. *Journal of Power Sources*, 65(1–2), 53–59.

24. Andreas, H. A. (2015). Self-discharge in electrochemical capacitors: A perspective article. *Journal of The Electrochemical Society*, 162(5), A5047.

25. Liu, K., Yu, C., Guo, W., Ni, L., Yu, J., Xie, Y., & Qiu, J. (2021). Recent research advances of self-discharge in supercapacitors: Mechanisms and suppressing strategies. *Journal of Energy Chemistry*, 58, 94–109.

26. Rizoug, N., Bartholomeus, P., & Le Moigne, P. (2012). Study of the ageing process of a supercapacitor module using direct method of characterization. *IEEE Transactions on Energy Conversion*, 27(2), 220–228.

27. Yu, A., Chabot, V., & Zhang, J. (2013). *Electrochemical Supercapacitors for Energy Storage and Delivery: Fundamentals and Applications* (p. 383). Taylor & Francis.

28. Varakin, I. N., Stepanov, A. B., & Menukhov, V. V. (1997). *Capacitor with a Double Electrical Layer*, Patent, Ref No: WO 97/07518/A1/.

29. Arshid N., Khalid, M., & Grace, A. N. (2021). *Advances in Supercapacitor and Supercapattery*. Elsevier.

30. Yu, L., & Chen, G. Z. (2019). Ionic liquid-based electrolytes for supercapacitor and supercapattery. *Frontiers in Chemistry*, 7, 272.

31. Tang, N., You, H., Li, M., Chen, G. Z., & Zhang, L. (2018). Cross-linked Ni (OH) 2/CuCo 2 S 4/Ni networks as binder-free electrodes for high performance supercapatteries. *Nanoscale*, 10(44), 20526–20532.

32. Yu, L., & Chen, G. Z. (2020). Supercapatteries as high-performance electrochemical energy storage devices. *Electrochemical Energy Reviews*, 3(2), 271–285.

33. Liu, H., Hu, R., Qi, J., Sui, Y., He, Y., Meng, Q., ... & Zhao, Y. (2020). A facile method for synthesizing NiS nanoflower grown on MXene (Ti3C2Tx) as positive electrodes for "supercapattery". *Electrochimica Acta*, 353, 136526.

34. Kaipannan, S., & Marappan, S. (2019). Fabrication of 9.6 V high-performance asymmetric supercapacitors stack based on nickel hexacyanoferrate-derived Ni (OH)$_2$ nanosheets and bio-derived activated carbon. *Scientific Reports*, 9(1), 1–14.

35. Oltean, I. D., Matoi, A. M., & Helerea, E. (2010, May). A supercapacitor stack-design and characteristics. In 2010 12th International Conference on Optimization of Electrical and Electronic Equipment (pp. 214–219). IEEE.

36. Ibanez, F., Vadillo, J., Echeverria, J. M., & Fontan, L. (2013, October). Design methodology of a balancing network for supercapacitors. In *IEEE PES ISGT Europe* 2013 (pp. 1–5). IEEE.

37. Liu, X., Wu, T., Dai, Z., Tao, K., Shi, Y., Peng, C., ... & Chen, G. Z. (2016). Bipolarly stacked electrolyser for energy and space efficient fabrication of supercapacitor electrodes. *Journal of Power Sources*, 307, 208–213.

38. Hansson, J., Li, Q., Smith, A., Zakaria, I., Nilsson, T., Nylander, A., ... & Enoksson, P. (2020). Bipolar electrochemical capacitors using double-sided carbon nanotubes on graphite electrodes. *Journal of Power Sources*, 451, 227765.

39. Habib, A. A., Hasan, M. K., Mahmud, M., Motakabber, S. M. A., Ibrahimya, M. I., & Islam, S. (2021). A review: Energy storage system and balancing circuits for electric vehicle application. *IET Power Electronics*, 14(1), 1–13.

40. Qu, Y., Zhu, J., Hu, J., & Holliday, B. (2013, June). *Overview of supercapacitor cell voltage balancing methods for an electric vehicle*. In 2013 IEEE ECCE Asia Downunder (pp. 810–814). IEEE.

41. Lee, Y., Jeon, S., Lee, H., & Bae, S. (2016). Comparison on cell balancing methods for energy storage applications. *Indian Journal of Science and Technology*, 9(17), 92316.

42. Diab, Y., Venet, P., & Rojat, G. (2006, November). *Comparison of the different circuits used for balancing the voltage of supercapacitors: Studying performance and lifetime of supercapacitors*. In ESSCAP'06 (pp. on-CD).

8 Conventional Applications of Supercapacitors

Gopakumar G. and S. Anas

CONTENTS

8.1 INTRODUCTION

Research on new, reliable, renewable, and sustainable sources of energy has paved the way for the design and development of novel energy conversion and storage devices. The energy storage, conversion, and utilization based on energy-efficient technologies such as windmills, solar cells, and fuel cells, have become the need of the day in meeting the basic requirements of man in modern life. Many of these renewable and sustainable energy-efficient technologies mainly rely on environment-friendly energy storage devices like batteries and supercapacitors (SCs) [1–3]. However, the advanced energy storage technologies based on batteries like lithium-ion, lithium-sulphur, aluminium-ion, metal-air, etc., developed so far lack high power densities [4, 5]. SCs, on the other hand, function as reliable and efficient energy storage devices that deliver high power when compared to batteries. SCs are thus emerging as promising electrochemical energy storage devices, which cater to the demands of global energy consumption [6, 7].

DOI: 10.1201/9781003258384-8

SCs are electrochemical capacitors that utilize electrical double layer and pseudocapacitance to attain higher order capacitance [8, 9]. Fast charging and discharging, long cycle life, high power density, safe operation, easy maintenance, and zero pollution are some of the typical advantages of SCs [10, 11]. They are also capable of repeated charging and discharging without any deterioration [12]. Recently, nanocomposites have been introduced into the synthesis of SCs to achieve increased power density and energy density simultaneously. The performance of nanocomposite electrodes, which is the sum of performances of individual components, is preferred to that of capacitors developed from the individual counterparts [13, 14]. Reasonable selection and design of materials is critical in improving the performance of SCs.

SCs have earned their significance in applications to power electric vehicles, electronic systems, smart devices, etc., requiring electrical energy for operation, by increasing their reliability and energy efficiency [15–17]. The introduction of hybrid electric vehicles has increased the potential applications of SCs further, given their extended energy support for the management of such vehicles. Many intelligent electronic systems have emerged in the industry, such as smart sensors, AI-based systems, robots, smart home technologies, smartphones, and allied microsystems. Matching energy storage devices are needed for integrating these microelectronic systems into the devices for various applications [18–22].

Industrial system managers are in continuous search of fool-proof energy storage systems. SCs had a huge impact on energy-related issues on applications like traction drives, power grids, consumer electronics, renewable energy, etc., and they emerged as a solution for most of the troublesome systems [23–25]. The uninterrupted growth in the field of portable, as well as wearable, electronic devices, has stimulated the need for safe and reliable power supply units. Successful advancements and wider commercialization of SCs can bring about enhanced market considerations. SCs have a broad prospect in future applications and some of the advanced applications of SCs in load levelling, regenerative braking, cranes, lifts, and trucks, and portable electronic systems are addressed in detail in this chapter.

8.2 LOAD LEVELLING

8.2.1 INTRODUCTION

An uninterrupted and frequent power supply has become a need for all and it has become an essential minimum requirement of modern life. All our energy distribution systems should be free from frequent voltage fluctuations and sudden load changes. SCs have a great role in smoothening the load pattern in power distribution systems by decreasing on-peak and increasing off-peak loads. This is generally referred to as 'load levelling' [26–32]. Many reliable and cost-effective power supplies now face threats from power demand fluctuations. Renewable energy resource-based distributed power generation often results in unpredictable and stochastic power supply due to accelerated weather change and unexpected power outages [32–42]. For example, during rainy seasons, solar photovoltaic (PV) energy production will be greatly reduced. This discrepancy in energy production due to sudden climate change may cause power quality problems. When injected into the grid, it may further result in power system instability. Since the SC-based energy storage systems offer power quality, power and voltage smoothing, peak shaving, and load-levelling benefits, they can be successfully integrated into these power distribution systems, for better efficiency, reliability, and power grid stability [43–51].

8.2.2 SUPERCAPACITORS IN LOAD-LEVELLING APPLICATIONS

SCs or electric double-layer capacitors are very attractive for load-levelling applications mainly because of three reasons. (1) The power densities of these SCs are approximately ten times higher than those of normal electrochemical batteries. (2) Compared with conventional aluminium electrolytic capacitors, the energy density of these SCs are approximately 100 times higher. (3) SCs have an extremely large number of charge/discharge cycles (~106) [52–55]. Because of these reasons, SCs are advantageously employed as storage devices for applications where high power is required for short time. The widespread applications of SCs include regenerative braking in hybrid electric vehicles, power smoothing in elevator applications and ropeways, and voltage sag compensation in unified power quality controllers. SCs can also be used as storage devices that could be applied as an energy buffer to the electric power system of overhead cranes (OCs). SCs filter higher frequencies to smooth Variable Renewable Energy Source (VRES) generation and they provide low voltage ride-through and remove the need for a DC/AC converter [56–61]. One of the biggest advantages of SCs, when compared to other energy storage technologies, is their fast response and high cyclability. If compared with other Electrical Energy Storage (EES) technologies like compressed air energy storage systems, pumped hydro storage, and batteries, it exhibits better cycle life. It can mitigate many problems associated with distributed VRES [60, 61]. Every distributed EES system requires low maintenance and has a long life to decrease costs associated with replacements, maintenance, and operations. Since it qualifies all the desired criteria, the SCs are considered as reliable and acceptable systems for any EES applications.

8.2.3 HYBRID SUPERCAPACITORS IN LOAD-LEVELLING APPLICATIONS

Hybrid SCs are energy-efficient SCs that can deliver or receive energy based on the power requirements. All our available energy storage systems can be integrated with SCs to prepare the 'hybrid capacitors'. One of the most common hybrid SC assemblies is the battery-SC-based hybrid system on which the SC functions as the load-levelling device for the batteries. Here, the SCs store the surplus energy from the battery during low load demand periods, and they provide the required extra current during high load demand periods [60–62]. The hybrid system offers almost 100% cycle efficiency (ratio of energy output to input), which makes them suitable for energy storage in a situation where frequent charge/discharge occurs. Such an assembly of SC-battery-based hybrid systems can meet the high peak-to-average power demand in electric vehicles. Here, it functions by dividing the overall power demand into high and low-frequency components. The high-frequency component will be taken care of by the SC and the low-frequency part will be supplied by the batteries. At a given point in time, the power provided by the hybrid system will be stable and the output voltage will be equal to the overall power demand.

All modern portable electronic devices which suffer from a highly fluctuating load profile now rely on the battery-SC hybrid systems which stabilize the energy loss due to the rate capacity effect. The hybrid system which is constructed in parallel configurations generally contains a large SC to minimize the batteries' peak discharging current and to relieve the effect of rate capacity on batteries. The overall energy density must be compromised in cases where the inclusion of a large SC degrades the overall energy density. However, in constant-current regulator-based hybrid systems, isolating the battery from SCs, relieves the rate capacity effect even with the use of smaller SCs [62–65].

8.3 REGENERATIVE BRAKING

8.3.1 INTRODUCTION

Braking energy is one of the most readily available forms of energy in vehicles, which is wasted most of the time. Due to friction losses, this energy is wasted in the form of heat in normal vehicles. But in electric vehicles, which make use of regenerative braking energy technologies, the kinetic energy generated by the motor during the braking process is utilized. The kinetic energy associated with the braking process can be converted and stored with the help of energy storage devices. Here, the motor functions as a generator, and the available kinetic energy is harvested. The harvested energy is stored with the help of necessary switching and energy storage systems. The stored energy can be used again for charging the vehicle's battery or can be stored in a capacitor or both. The stored energy can be converted again to kinetic energy with the help of controllers and can be used for starting, or acceleration of the electric vehicles [66, 67].

So far, many energy storing devices are used for storing the braking energy. Among them, the first used are mainly resistor energy dissipation types. Even though these types are reliable in operation and have the lowest cost, they have been phased out mainly because of the pollution and the associated energy wastage. The next used is the energy feedback-type devices which possess advantages like effective utilization of the braking energy, but they are also rejected because of the difficulties associated with coordinating the energy feedback system with the power grid. The battery-type storage devices which were developed later resolved almost all the issues associated with the earlier devices and received special attention in the market. But, due to the low power density, short lifetime, and high energy device, the batteries are not found to be as successful as expected. For instance, the braking feedback transient current in an electric vehicle that runs at a very high speed can reach up to 200 A, which can cause damage to the batteries even with lithium-ion batteries [55, 67, 68]. The recent scientific innovations in the field have proved that SCs can be advantageously employed for this purpose. Since they possess high power density, the braking energy can be quickly converted from kinetic energy to electric energy. Now, SC-based energy storage devices have become the research hotspot. They exhibited specific advantages like good compatibility, rapid charge/discharge rates, long cycle life, and high power density ($P > 10$ kW kg^{-1}), which ultimately improved energy savings and extended the driving range of electric vehicles [55, 67–71].

8.3.2 SUPERCAPACITORS IN REGENERATIVE BRAKING APPLICATIONS

SC-based regenerative braking technologies which are utilized in modern electric vehicles function with the following principle. When the brake is applied in electric vehicles, the required brake torque is produced by an in-wheel motor. According to the braking current intensity, the recovery direction of braking energy will be controlled by DC/DC controller. If the braking current is less than the charging current threshold, the braking energy is stored in the battery pack directly. Otherwise, the braking energy is stored in the SC [67–73].

SCs have been used quite extensively for regenerative braking applications in battery- or hybrid-powered cars, trains, lifts, etc. [74–84]. They are used as independent energy sources in vehicles particularly for short-distance electric vehicles such as city buses, terminal trucks, and tunnel trucks [85–88]. Recent developments in electric vehicles highlight the composite structures based on lithium-ion batteries and SCs to increase energy efficiency and to improve vehicle safety. Vehicles using normal batteries suffer a short running distance in one battery charge cycle and this is also limited to the vehicle size and load-carrying capability of the vehicle [81, 89–93]. Since the regenerative braking energy can be treated as an alternative energy source, this energy can be effectively integrated for charging the batteries with the help of a SC. Recent research developments have proved that graphene-based SCs are excellent energy storage systems that can store regenerative energy during braking/deceleration modes. They function as an auxiliary power source to drive electric vehicles. The studies highlight the potential use of a composite/hybrid SC for ideal electrical energy storage and conversion systems for next-generation electric vehicles [94–96].

8.4 CRANES, LIFTS, AND TRUCKS

8.4.1 INTRODUCTION

Cranes, lifts (elevators), and trucks are the equipment of importance in commercial and industrial processes. They are used for heavy operations like loading, earth-moving, lifting, transportation, etc. Conventionally, these are powered by engines working on the combustion of fossil fuels. Electrochemical batteries are used for starting engines and internal combustion engines are used for the power of propulsion [97]. But the over-dependence of this heavy machinery on batteries and fossil fuels creates pollution as well as greenhouse gas emission issues [98]. Hence, major improvements in the design and operation of cranes, lifts, and trucks are now focusing on the reduction of energy consumption. Environmental aspects, as well as energy problems, have become a pressing need in the context of the promotion of renewable energies and energy saving. The use of storage devices for potential savings is considered remarkable for energy recovery and global efficiency. Here, SCs are becoming an interesting choice for the said heavy equipment based on the application considered [99].

The high-power density of SCs makes them attractive for uses where high power is required for short time. SCs find general applications in cranes, lifts, and trucks. It can be used for;

- Short term energy storage
- Protection of batteries from power fluctuations
- The key component of regenerative braking
- Transmission and lighting
- Burst mode power delivery for ignition and engine cranking

In the case of internal combustion (IC) engines, a SC functions with starters to provide immediate energy bursts for ignition and gives the battery a longer life. For example, hybrid SCs can be linked to IC engine cranking by providing rapid energy bursts. The compatible SC and starter motor lower the cost of maintenance than that of lead-acid batteries [100]. The usage of SCs for transportation is motivating and improves energy efficiency to overcome the wastage of energy as heat. Here, SCs will function as *regeneration devices* that recover the energy released by the repetitive and constant movements of machines [101]. Hence, we can find SCs in cranes, trucks, and lifts or in the braking systems of hybrid or electric vehicles. This innovation can reduce the consumption of fuel as well as the emission of carbon dioxide. The transportation industry alone is responsible for about 30% of global emissions of greenhouse gases, especially from cargo transport [102]. This means that any energy efficiency enhancement in cargo transportation will inevitably have a global-level impact. Hence, SC-based energy storage systems are promising innovations to improve the efficiency of transportation.

8.4.2 Supercapacitors in Cranes

Overhead cranes are a major component of cargo handling apparatus found in industrial and commercial applications. The operating mechanism of overhead cranes has been a topic of discussion among entrepreneurs as well as researchers [103, 104]. The combustion of fossil fuel is the prime source of electrical energy for cranes working in all manufacturing and logistics units [105]. The burning induces the emission of greenhouse gases and air pollution. This necessitates improvement in crane operation strategies focused on energy efficiency.

The main area of the energy flow in a crane is the 'hoist'. A raised payload buffers a notable amount of potential energy. During the lowering of a payload, the energy of gravity is ineffectually squandered as heat energy in the brake or brake resistor [106]. SCs based on double-layer-electrolyte ability can be effectively used for storing such energy after converting it into electrical energy. The energy brought about when a container is lowered is usually burned in the braking resistors. The excess energy can be detained and reused to help lift the containers with the presence of a SC-powered system [107]. This solution for cranes decreases energy consumption as well as fuel usage.

Cost-effective SCs are mainly useful in loading and unloading heavy cargo containers at seaports. SCs have certain definite functions during loading operations and unloading as well. One important function is to capture the energy that may otherwise be wasted as heat during repeated up-and-down movements. Lifting a load generally needs a arge amount of energy. But some amount of energy can be saved during lowering which can be used for improving energy efficiency [108, 109]. The second important function is to permit a size reduction in the primary source of power which is usually an IC engine. This is called peak shaving. The engine is designed to operate at average levels of power by sizing the SC [110]. This can reduce gas emissions and thus improve the quality of air.

SCs can be used alone or in combination with batteries or any other storage system to offer enhanced power efficiency and improved cycle life for applications in heavy vehicles, cranes, etc. [111, 112]. Another application of SCs is related to starters by providing pulse power to start the IC engine, which is now reduced due to the discouraging of fossil fuel power.

8.4.3 SUPERCAPACITORS IN TRUCKS AND LIFTS

SCs are used in heavy vehicles, traction lifts, ropeways, and elevators for engine starting, power modulation, and regenerative braking [113–115]. Dangers and health concerns make all-electric forklift trucks very common. These vehicles may have a battery or a fuel cell for energy storage. In addition, SCs will act as power sources supporting operations and as a device for recovering braking energy [116]. Usually, secondary cells are used for engine cranking applications in heavy vehicles. But SCs with the capacity to produce rapid energy bursts can be used to start large diesel engines effectively during difficult conditions like low temperature [117–119]. Also, high power density and rapid charging and discharging of SCs make them suitable for capturing kinetic energy as a vehicle slows down [89, 120, 121]. SCs can release the energy for bursts of acceleration, especially during the gear changing of mechanical transmission by bridging up the gap in the delivery of power. SCs can discharge during acceleration and recharge during braking [122]. Trucks use batteries and fuel cells as primary energy storage units, while SCs are used for regenerative braking purposes. SCs, when paired with batteries, can extend the life of batteries in electrified commercial vehicles such as buses or trucks.

For cranes and lifts, SCs can provide emergency power during unexpected power failures. This may reduce the risk of sudden stoppage and consequent accidents. Also, SCs can be employed for load-levelling applications where fluctuations of power are expected. They can recover energy during the lowering of the cage in cranes and lifts, which may otherwise be lost as dissipated heat. SCs act as power-smoothing devices in elevators and ropeways. The braking energy of lifts can be stored in SCs and the lift makes use of this energy, consequently reducing the consumption [123,

124]. This can be achieved without much alteration in the design and control system of the lift. A system for the recovery of kinetic energy is an ideal application of a SC.

Carbon-type SCs have been successfully used in combination with lead-acid batteries in trucks [125]. SCs integrated with fuel cell-powered vehicles have advantages like functioning without interface electronics. SCs are a promising energy storage technology for refuse trucks experiencing thousands of start/stop cycles in a day. The capacity of high power, exceptional efficiency, and the ability to operate in challenging environments, low maintenance, decreasing price, safety, and long lifetime of SCs make them a wonderful option for potential applications in the field of vehicles, passenger, and traction lifts.

8.5 CONSUMER ELECTRONICS

The durability, high output power, and enhanced energy density of SCs have established the possibility of SCs being used to power portable electronic equipment. Among the various energy storage systems, SCs with surface charge storage mechanisms are capable of providing high power density within a short time [9, 126–128]. SCs, unlike conventional capacitors, make use of the total capacitance contributed by electrostatic double-layer capacitance and electrochemical pseudocapacitance. They receive and deliver charges much faster than batteries and contribute greatly towards the replacement of rechargeable batteries. A detailed discussion of the application of SCs in portable electrodes is beyond the scope of this chapter and is provided in Chapter 9.

8.6 SUMMARY AND OUTLOOK

The abstract idea and success of SCs in the fields of load levelling, regenerative braking, and portable electronics have attracted the attention of the world of science and technology since the beginning due to the outstandingly high capacitance and unlimited cycle life of charging and discharging shown by SCs. The capacity of incredible energy storage, ultra-fast charging, and rapid discharge helped SCs to render such exceptional power-related applications. Current technologies are far more advanced than their conception in the 19th century, with the potential for advanced energy storage technologies demanded by the present. Innovative technologies based on SCs are already used by giant electronics companies for transportation and energy solutions. At present, the applications of SCs are confined to utility vehicles, power banking, rail power systems, hybrid vehicles as well as other automotive and grid stabilization, owing to their capacity to enhance the performance of the production systems and ability to improve the energy efficiency and reliability of electrical systems. Based on the capacities of SCs, enormous research activities are underway with a motive to improve the performance of various electrical systems engaged in industrial processes. In the future, SCs with increased energy density will be more preferred to batteries and other energy storage devices due to the demands of high power, time saving, green technology regulations, and clean energy requirements.

Energy storage systems, based on SCs or electric double-layer capacitors (EDLCs), are now becoming an essential tool for solving the issues of high-power technologies related to industry. This is the contribution of researchers working with power grids, electric traction, and power generation using renewable technologies. Integration of EDLC technology with energy storage systems may pave the way for futuristic applications including:

Electric traction in vehicles on road and rail
  Renewable energy generation including wind energy, tidal energy, and solar PV energy
  Power grid connection applications like grid regulations, micro-grids, frequency compensation, and voltage compensation

Research is getting closer towards the development of standalone SC batteries which take up only a fraction of the space of lithium-ion cells. They can be charged more quickly and can be recharged more than 25,000 times. The achievements so far are thrilling realizations of the present-day technologies which will be replaced with more feasible devices in near future. The concept of lightweight SCs with very high energy storage capacity (up to 500 F/g) made of graphene is an innovative technology to be explored for changing portable electronics as well as the total capacitor business. The most auspicious future may be the blending of a double-layer charging interface with fuel cells, batteries, or other conventional energy sources for use in hybrid electric vehicles. Continuous improvements in load-levelling systems will also have an impact on the harmony of the global community rather than for a commercial consumer.

We have witnessed great progress made by science and technology in the development of SCs in the last decade. The data from the most relevant publications of the last five years indicates the direction of progress of SCs towards cost-effective, safe, and sustainable devices with enhanced storage capacity. SCs can emerge with exciting results for solving issues regarding environmental hazards, portability, increased fuel prices, and intrinsic safety. New developments like smart homes and 5G technology increase the need for ideal energy sources where SCs can offer excellent solutions. The successful creation of high-performance SCs will empower researchers to produce eco-friendly portable systems of extended durability.

Various discussions regarding novel systems for energy storage indicate that a combination of batteries and SCs will be dominant in the future. These hybrid systems can deliver performance greater than the sum of individual operations. This is deemed to be beneficial to high-power electric motors with plug-in hybrids, where the limiting of the weight and size of batteries is almost impossible. The successful development and future marketing of electrical and mechanical components of hybrid vehicles are very dependent on the quality of energy storage systems designed in the future. Asymmetric SCs are also definitely research-directed towards the development of environment-friendly systems at affordable prices. Future research work may focus on developing and optimizing environment-friendly procedures involving simplified doping or activation of natural precursors.

Researchers can start thinking of seamless integration of self-powered appliances like sensors and circuits with flexible and durable materials. Devices working on sweat power can be successfully integrated into modern technologies like robotics, IoT, augmented reality, virtual reality, and smart home technologies enabling the establishment of a better link between humans and technology. Now, people are paying attention to consuming energy that is harmless to the environment. The rapid development of consumer electronics also demands portable power supply sources with higher-order capacities. This could drive the development of SCs leading to broad market perspectives.

REFERENCES

1. Xing, L., Jihong, W., Mark, D., & Jonathan, C., (2015) Overview of current development in electrical energy storage technologies and the application potential in power system operation, *Applied Energy*, 137, 511–536.
2. Olabi, A.G., Onumaegbu, C., Wilberforce, T., Ramadan, M., Abdelkareem, M.A., & Abdul Hai, A., (2021) Critical review of energy storage systems, In *Energy*, Elsevier, 214, 118987.
3. Mustafizur, R., Abayomi, O.O., Eskinder, G., & Amit, K., (2020) Assessment of energy storage technologies: A review, *Energy Conversion and Management*, 223, 113295.
4. Bruce, D., Haresh, K., & Jean, M.T., (2011) Electrical energy storage for the grid: A battery of choices, *Science*, 334, 928–935.
5. Etacheri, V., Marom, R., Elazari, R., Salitra, G., & Aurbach, D., (2011) Challenges in the development of advanced Li-ion batteries: A review, *Energy & Environmental Science*, 4(9), 3243–3262.
6. Kamal, K., (2020) *Handbook of Nanocomposite Supercapacitor Materials II*. Springer.
7. George, Z.C., (2017) Supercapacitor and supercapattery as emerging electrochemical energy stores, *International Materials Reviews*, 62(4), 173–202.
8. Hall, P.J., Mirzaeian, M., Fletcher, S.I., Sillars, F.B., Rennie, A.J.R., Shitta-Bey, G.O., Wilson, G., Cruden, A., & Carter, R., (2010) Energy storage in electrochemical capacitors: Designing functional materials to improve performance, *Energy & Environmental Science*, 3(9), 1238–1251.
9. Wang, G., Zhang, L., & Zhang, J., (2012) A review of electrode materials for electrochemical supercapacitors, *Chemical Society Reviews*, 41(2), 797–828.
10. Ioannis, H., Andreas, P., & Venizelos, E., (2009) Overview of current and future energy storage technologies for electric power applications, *Renewable and Sustainable Energy Reviews*, 13(6–7), 1513–1522.
11. Zhibin, Z., Mohamed, B., Jean F, Franck S., & Tianhao T., (2013) A review of energy storage technologies for marine current energy systems, *Renewable and Sustainable Energy Reviews*, 18, 390–400.
12. Maximilian, K., Julia, K., & Dirk U.S., (2010) Modelling the effects of charge redistribution during self-discharge of supercapacitors, *Electrochimica Acta*, 55(25), 7516–7523.
13. Elmira, P., & Reza, T.M., (2021) A novel ternary Fe3O4@Fc-GO/PANI nanocomposite for outstanding supercapacitor performance, *Electrochimica Acta*, 383, 138296.
14. Thibeorchews, P., Biny, R.W., Gautam, C.R., Rajangam, I., & Sujin, P.J., (2018) Synthesis and enhanced electrochemical performance of PANI/Fe3O4 nanocomposite as supercapacitor electrode, *Journal of Alloys and Compounds*, 757, 466–475.
15. Pandolfo, T., Ruiz, V., Sivakkumar,S., & Nerkar, J., (2013) *General Properties of Electrochemical Capacitors in Supercapacitors*, Chap. II, Wiley, 69–109.

16. Simon, P., Taberna, P.L., & Béguin, F., (2013) *Electrical Double-Layer Capacitors and Carbons for EDLCs in Supercapacitors* Chap. IV, Wiley, 131–165.

17. Beaudin, M., Zareipour, H., Schellenberg, A., & Rosehart, W., (2014) Energy storage for mitigating the variability of renewable electricity sources. Energy storage smart grids plan, *Operation for Renewable and Variable Energy Resources*, 14(4), 302–314.

18. González, A., Goikolea, E., Barrena, J.A. & Mysyk, R., (2016) Review on supercapacitors: Technologies and materials, *Renewable and Sustainable Energy Reviews*, 58, 1189–1206.

19. Chen, D., Lou, Z., Jiang, K., & Shen, G., (2018) Device configurations and future prospects of flexible/stretchable lithium-ion batteries, *Advanced Functional Materials*, 28, 1805596.

20. Wang, Y., & Xia, Y., (2013) Recent progress in supercapacitors: from materials design to system construction, *Advanced Materials Technologies*, 25, 5336–5342.

21. Kyeremateng, N.A., Brousse, T., & Pech, D., (2017) Microsupercapacitors as miniaturized energy-storage components for on-chip electronics, *Nature Nanotechnology*, 12, 7–15.

22. García Núñez, C., Manjakkal, L., & Dahiya, R., (2019) Energy autonomous electronic skin, *npj Flexible Electronics* 3, Article number: 1.

23. Navarro, G., Torres, J., Blanco, M., Nájera, J., Santos-Herran, M., & Lafoz, M., (2021) Present and future of supercapacitor technology applied to powertrains, renewable generation and grid connection applications, *Energies*, 14(11), 3060.

24. Ruddell, S., Madawala, U.K., & Thrimawithana, D.J., (2020) A wireless EV charging topology with integrated energy storage, in *IEEE Transactions on Power Electronics*, 35(9), 8965–8972.

25. Passalacqua, M., Carpita, M., Gavin, S., Marchesoni, M., Repetto, M., and Vaccaro, L., & Wasterlain, S., (2019) Supercapacitor storage sizing analysis for a series hybrid vehicle, *Energies*, 12, (9) 1759.

26. Hemmati, R., & Saboori, H., (2016) Short-term bulk energy storage system scheduling for load levelling in unit commitment: Modeling, optimization, and sensitivity analysis, *Journal of Advanced Research*, 7(3), 360–372.

27. Kapsali, M., & Kaldellis, J.K., (2010) Combining hydro and variable wind power generation by means of pumped-storage under economically viable terms, *Applied Energy*, 87(11), 3475–85.

28. Barzin, R., Chen, J.J., Young, B.R., & Farid, M.M., (2015) Peak load shifting with energy storage and price-based control system, *Energy*, 92, 505–14.

29. Li, Z., Guo, Q., Sun, H., & Wang, J., (2015) Storage-like devices in load leveling: Complementarity constraints and a new and exact relaxation method, *Applied Energy*, 151, 13–22.

30. Han, X., Ji, T., Zhao, Z., & Zhang, H., (2015) Economic evaluation of batteries planning in energy storage power stations for load shifting, *Renewable Energy*,78, 643–647.

31. Zhuk, A., Zeigarnik, Y., Buzoverov, E., & Sheindlin, A., (2016) Managing peak loads in energy grids: Comparative economic analysis, *Energy Policy*, 88, 39–44.

32. Kloess, M., & Zach, K., (2014) Bulk electricity storage technologies for load-leveling operation: An economic assessment for the Austrian and German power market, *International Journal of Electrical Power and Energy Systems*, 59, 111–22.

33. Younghyun, K., Jason, K., Qing, X., Yanzhi, W., Naehyuck, C., & Massoud, P., (2014) A scalable and flexible hybrid energy storage system design and implementation, *Journal of Power Sources*, 255, 410–422.

34. *Electricity Supply-demand Outlook & Measures for the Summer of FY2013*, Agency for Natural Resources and Energy Ministry of Economy, Trade and Industry, Japan, 2013.

35. Nguyen, F., & Stridbaek, U., (2007) *Tackling Investment Challenges in Power Generation*, Tech. rep., International Energy Agency.

36. Wiehagen, J., & Harrell, D., (2001) *Review of Residential Electrical Energy Use Data,* Tech. rep., NAHB Research Center, Inc.,

37. Om, K., & Sathans, S., (2018) An updated review of energy storage systems: Classification and applications in distributed generation power systems incorporating renewable energy resources, *International Journal of Energy Research,* 43(12), 1–40.

38. Maria C.A., Paterakis, F., Christina P., Christos M., M. Darwish, & Marouchos C.C, (2018) *Supercapacitor Application for PV Power Smoothing Fluctuations in Solar PV In 2018 53rd International Universities Power Engineering Conference, 978-1-5386-2910-9/18/© IEEE.*

39. Thekaekara, M.P., (1976) Solar radiation measurement: Techniques and instrumentation, *Journal of Solar Energy,* 18(4), 309–325.

40. Parra, D., Walker, G.S., & M. Gillott, (2014) Modeling of PV generation, battery and hydrogen storage to investigate the benefits of energy storage for single dwelling, *Sustainable Cities and Society,*10, 1–10.

41. Beaudin, M., Zareipour, H., Schellenberglabe, A., & Rosehart, W., (2010) Energy storage for mitigating the variability of renewable electricity sources: An updated review, *Energy for Sustainable Development,*14(4) 302–314.

42. Senjyu, T., Datta, M., Yona, A., & Funabashi, T., (2008) PV output power fluctuations smoothing and optimum capacity of energy storage system for PV power generator, *Icrepq.Com,* 1(6) 3–7.

43. Seo, H.R., Kim, G.H., Kim, S.Y., Kim, N., Lee, H.G., Hwang, C., Park, M., &. Yu, I.K (2010) Power quality control strategy for gridconnected renewable energy sources using PV array and supercapacitor, In International Conference on Electrical Machines and Systems, 437–441.

44. Wang, L., Vo, Q.S., & Prokhorov, A.V., (2018) Stability improvement of a multimachine power system connected with a large- scale hybrid wind-photovoltaic farm using a supercapacitor, *IEEE Transactions on Industry Applications,* 54(1), 50–60.

45. Ghiassi-Farrokhfal, Y., Keshav, S., & Rosenberg, C., (2015) Towards a realistic performance analysis of storage systems in smart grids, *IEEE Transactions on Smart Grid,* 6(1), 402–410.

46. Zahedi, A., (2011) Maximizing solar PV energy penetration using energy storage technology, *Renewable and Sustainable Energy Reviews,*15(1), 866–870.

47. Carnegie, R., Gotham, D., Nderitu, D., & Preckel, P.V., (2013) Utility scale energy storage systems: Benefits, applications, and technologies, *State Utility Forecasting Group. Purdue University,* 1.

48. Argyrou, M.C., Christodoulides, P., & Kalogirou, S.A., (2018) Energy storage for electricity generation and related processes: Technologies appraisal and grid scale applications, *Renewable and Sustainable Energy Reviews,* 94, 804–821.

49. Molina, M.G. (2017) Energy storage and power electronics technologies: A strong combination to empower the transformation to the smart grid, *Proceedings of the IEEE,* 105(11), 2191–2219.

50. Naish, C., McCubbin, I., Edberg, O., & M. Harfoot (2008) Outlook of energy storage technologies, *Policy Department Economic and Scientific Policy February,* 1–57.

51. Schoenung, S.M., (2001) Characteristics and technologies for long-vs. short-term energy storage: A study by the DOE energy storage systems program, *United States Dep. Energy.*

52. Pasquali M., Tricoli P., Member, IEEE, & C. Villante., (2011) Testing Methodologies of Supercapacitors for Load-leveling Purposes in Industrial Applications, In *2011 International Conference on Clean Electrical Power (ICCEP)* (pp. 395–399). IEEE. *978-1-4244-8930-5/11/IEEE.*

53. Winter, M., & Brodd, R.J., (2004) What are batteries, fuel cells, and supercapacitors? *Chemical Reviews,* 104(10), 4245–4269.

54. Sharma, P., & Bhatti, T.S., (2010) A review on electrochemical double-layer capacitors, *Energ Conver Manage*, 51(12), 2901–2912.
55. Simon, P., Gogotsi, Y., & Dunn, B., (2014) Where do batteries end and supercapacitors begin? *Science*, 343, 1210–1211.
56. Li, W., Joos G., & Abbey C., (2006) Attenuation of wind power fluctuations in wind turbine generators using a DC bus capacitor based filtering control scheme, *Proceedings of Industry Applications Conference*, 41st IAS Annual Meeting. Conference Record of the 2006 IEEE; 1, 216–21.
57. Marc, B., Hamidreza, Z., Anthony, S., & William, R., (2010) Energy storage for mitigating the variability of renewable electricity sources: An updated review, *Energy for Sustainable Development*, 14, 302–314.
58. Enslin, J., & Heskes, P., (2004) Harmonic interaction between a large number of distributed power inverters and the distribution network, *IEEE Transactions on Power Electronics*, 19(6), 1586–93.
59. Weisser, D., & Garcia, R.S., (2005) Instantaneous wind energy penetration in isolated electricity grids: Concepts and review, *Renewable Energy*, 30(8), 1299–308.
60. Ngamroo, I., Cuk, S.A., Dechanupaprittha, S., & Mitani,Y., (2009) Power oscillation suppression by robust SMES in power system with large wind power penetration, *Physica C*, 469(1), 44–51.
61. Tande J. (2003) Grid integration of wind farms, *Wind Energy*, 6(3), 281–95.
62. Donghwa, S., Younghyun, K., Yanzhi, W., Naehyuck, C., & Massoud, P., (2012) Constant-current regulator-based battery-supercapacitor hybrid architecture for high-rate pulsed load applications, *Journal of Power Sources*, 205, 516–524.
63. Emadi, A., (2005) *Handbook of automotive power electronics and Motor Drives*. CRC Press.
64. Donghwa, S., Younghyun, K., Jaeam, S., Naehyuck, C., Yanzhi, W., & Massoud, P., (2010) *Battery-Supercapacitor Hybrid System for High-Rate Pulsed Load Applications*, report.
65. Garcia, F.S., Ferreira, A.A., & Pomilio, J.A., (2009) *Control Strategy for Battery-Ultracapacitor Hybrid Energy Storage System, Conference Paper*
66. Julius, P., & Dina, I.A., (2019) The role of supercapacitors in regenerative braking systems, *Energies*, 12, 2683.
67. Yutong, P., Yehui, Z., Xiaohao, Z., Gengyu, L., & Ce Z., (2017) Status Analysis of Regenerative Braking Energy Utilization Equipments in Urban Rail Transit, IEEE Transportation Electrification Conference and Expo, Asia-Pacific, ITEC Asia-Pacific.
68. Ren, G., Ma, G., & Cong, N., (2015) Review of electrical energy storage system for vehicular applications, *Renewable Sustainable Energy Reviews*, 41, 225–236.
69. Horn, M., MacLeod, J., Liu, M., WebbJ., & Motta, N., (2019) Supercapacitors: A new source of power for electric cars? *Economic Analysis and Policy*, 61, 93–103.
70. Xiong, T., Lee, W.S.V., Chen, L., Tan, T.L., Huang, X., & Xue, J., (2017) *Energy & Environmental Science*, 10, 2441.
71. Andrew, A., & Rached, D., (2017) Modeling and analysis of a regenerative braking system with a battery-supercapacitor, *Energy Storage*, 978–1-5090-5454-1/17/2017. IEEE.
72. Kim, T.H., Jung-Hyo, L., & Chung-Yuen W., (2014) Design and control methods of bidirectional DC-DC converter for the optimal DC link voltage of PMSM drive, *Journal of Electrical Engineering and Technology*, 9(6), 1944–1953.
73. Jiaqun, X., & Haotian, C., (2015) Regenerative brake of brushless DC motor for light electric vehicle, In 2015 18th International Conference on Electrical Machines and Systems (ICEMS), Pattaya, 1423–1428.
74. Ceuca, E., Tulbure, A., & Risteiu, M., (2010) The evaluation of regenerative braking energy, In IEEE 16th International Symposium for Design and Technology in Electronic Packaging (SIITME).

75. Siddharth, S., Kavita, B., & Amit, S., (2012) Energy saving through regenerative braking in diesel locomotive with super-capacitors, *International Journal on Emerging Technologies*, 3(2), 109–114.

76. Naseri, F., Farjah E., & Ghanbari, T., (2016) An efficient regenerative braking system based on battery/supercapacitor for electric, hybrid and plug-in hybrid electric vehicles with BLDC motor, In IEEE Transactions on Vehicular Technology, 99.

77. Ding, S., Cheng, M., Chao H., Guishu, Z., & Wei, W., (2013) An energy recovery system of regenerative braking based permanent magnet synchronous motor for electric vehicles, In 2013 International Conference on Electrical Machines and Systems (ICEMS), Busan, 280–284.

78. Sindhuja M., Karthikeyan K., Arunprasath, S., & Sang-Jae, K., (2021) High-power graphene supercapacitors for the effective storage of regenerative energy during the braking and deceleration process in electric vehicles, *Materials Chemistry Frontiers*, 5, 6200–6211.

79. Nomura, K., Nishihara, H., Kobayash, N., Asada T., & Kyotani, T., (2019) 4.4 V supercapacitors based on super-stable mesoporous carbon sheet made of edge-free grapheme walls, *Energy & Environmental Science*, 12, 1542–1549.

80. Schupbach, R.M., Balda, J.C., Zolot, M., & Kramer, B., (2003) IEEE 34th Annual Conference on Power Electronics Specialist, PESC '03., IEEE, 1, 88–93.

81. Kim, B.K.., Sy, S., Yu, A., & Zhang, J., (2015) *Handbook of Clean Energy Systems*, John Wiley & Sons, Ltd, Chichester, UK, 1–25.

82. Wang, D-W.W., Li, F., Liu, M., Lu, G.Q., & Cheng, H-M.M., (2008) 3D aperiodic hierarchical porous graphitic carbon material for high-rate electrochemical capacitive energy storage, *Angewandte Chemie International Edition*, 47, 373–376.

83. Dixon, J., Nakashima, I., Arcos, E.F., & Ortuzar, M., (2010) Electric vehicle using a combination of ultracapacitors and ZEBRA battery, *IEEE Trans. Ind. Electron.*, 57, 943–949.

84. Zou, Z., Cao, J., Cao, B., & Chen, W., (2014) Evaluation strategy of regenerative braking energy for supercapacitor vehicle, *ISA Transactions*, 55, 234–240.

85. Teymourfar, R., Asaei, B., Iman-Eini, H., & Fard, R.N., (2012) Stationary super-capacitor energy storage system to save regenerative braking energy in a metro line, *Energy Conversion and Management*, 56, 206–214.

86. Zupan, I., Sunde, V., Ban, Z., & Kruselj, D., (2021) Algorithm with temperature-dependent maximum charging current of a supercapacitor module in a tram regenerative braking system, *Journal of Energy Storage*, 36, 102378.

87. González-Gil, A., Palacin, R., & Batty, P., (2013) Sustainable urban rail systems: Strategies and technologies for optimal management of regenerative braking energy, *Energy Conversion and Management*, 75, 374–388.

88. Ceuca, E., Tulbure, A., & Risteiu, M., (2010) The evaluation of regenerative braking energy, In IEEE 16th International Symposium for Design and Technology in Electronic Packaging (SIITME), 65–68.

89. Zhang, H., Wang, Y., & Soon, P.L., (2014) Application of super capacitor in HEV regenerative braking system, *The Open Mechanical Engineering Journal*, 8, 581–586.

90. Massot-Campos, M., Montesinos-Miracle, D., Bergas-Jané, J., & Rufer, A., (2012) Multilevel modular DC/DC converter for regenerative braking using supercapacitors, *Journal of Energy and Power Engineering*, 6, 1131–1137.

91. Itani, K., De Bernardinis, A., Khatir, Z., & Jammal, A., (2017) Comparative analysis of two hybrid energy storage systems used in a two front wheel driven electric vehicle during extreme start-up and regenerative braking operations, *Energy Conversion and Management*, 144, 69–87.

92. Kouchachvili, L., Yaici, W., & Entchev, E., (2018) Hybrid battery/supercapacitor energy storage system for the electric vehicles, *J. Power Sources*, 374, 237–248.

93. Uddin, K., Moore, A.D., Barai, A., & Marco, J., (2016) The effects of high frequency current ripple on electric vehicle battery performance, *Applied Energy*, 178, 142–154.

94. Karandikar, P.B, Talange, D.B., Sarkar, A., Kumar, A., Singh, G.R., & Pal, R., (2011) Feasibility study and implementation of low cost regenerative braking scheme in motorized bicycle using supercapacitors, *Journal of Asian Electric Vehicles*, 9, 1497–1504.

95. Krishnamoorthy, K., Pazhamalai, P., Mariappan, V.K., Manoharan, S., Kesavan, D., & Kim, S.J., (2020) Two-dimensional siloxene–graphene heterostructure-based high-performance supercapacitor for capturing regenerative braking energy in electric vehicles, *Adv. Funct. Mater.*, 31, 2008422.

96. Naseri, F., Farjah, E., & Ghanbari, T., (2016) An efficient regenerative braking system based on battery/supercapacitor for electric, hybrid and plug-in hybrid electric vehicles with BLDC motor, *IEEE Trans. Veh. Technol.*, 66(5), 3724–3738.

97. Dell, R.M., Moseley, P.T., & Rand, D.A.J., (2014) Chapter 5 - *Progressive Electrification of Road Vehicles, Towards Sustainable Road Transport*, Academic Press, 157–192.

98. Gielen, D., Boshell, F., Saygin, D., Bazilian, M.D., Wagner, N., & Gorini, R., (2019) The role of renewable energy in the global energy transformation, *Energy Strategy Reviews*, 24, 38–50.

99. Forouzandeh, P., Kumaravel, V., & Pillai, S.C., (2020) Electrode materials for supercapacitors: A review of recent advances, *Catalysts*, 10(9), 969.

100. Park, D.W., Cañas, N.A., Schwan, M., Milow, B., Ratke, L., & Friedrich, K.A., (2016) A dual mesopore C-aerogel electrode for a high energy density supercapacitor, *Current Applied Physics*, 16(6), 658–664.

101. Wang, T., & Wang, Q., (2014) An energy-saving pressure-compensated hydraulic system with electrical approach, *IEEE/ASME Transactions*, 19, 570–578.

102. Shaheen, S.A., & Lipman, T.E., (2007) Reducing greenhouse emissions and fuel consumption: Sustainable approaches for surface transportation, *IATSS Research*, 31(1), 6–20.

103. Sikora, M., Szczybra, K., Wróbel, Ł., & Michalak, M., (2019) Monitoring and maintenance of a gantry based on a wireless system for measurement and analysis of the vibration level, *Eksploatacja i Niezawodnosc - Maintenance and Reliability*, 21(2), 341–350.

104. Zelic, A., Zuber, N., & Stostakov, R., (2018) Experimental determination of lateral forces causes by bridge crane skewing during travelling, *Eksploatacjai Niezawodnosc - Maintenance and Reliability*, 20(1), 90–99.

105. Honczarenko, J., & Berliński, A., (2011) Modelling energy consumption of transport processes in automated assembly systems, *Assembly Technology and Automation*, 4, 49–52.

106. Kosucki, A., Stawiński, L., Malenta, P., Zaczyński, J., & Skowrońska, J., (2020) Energy consumption and energy efficiency improvement of overhead crane's mechanisms, *Eksploatacja i Niezawodnosc - Maintenance and Reliability*, 22, 323–330.

107. Conte, M., Genovese, A., Ortenzi, F., & Vellucci, F., (2014) Hybrid battery-supercapacitor storage for an electric forklift: A life-cycle cost assessment, *Journal of Applied Electrochemistry*, 44, 1–10.

108. Parise, G., Parise, L., Malerba, A., Pepe, F.M., Honorati, A., & Chavdarian, P.B., (2017) Comprehensive peak-shaving solutions for port cranes, In *IEEE Transactions on Industry Applications*, 53(3), 1799–1806.

109. Parise, G., Honorati, A., Parise, L., & Martiranon, L., (2015) Near zero energy load systems: The special case of port cranes, *Proc. IEEE/IAS 51st Ind. Comm. Power Syst. Techn. Conf.*, 1–6.

110. Ferreira, K., dos Santos, W.M., & Rueda Medina, A.C., (2019) Sizing of supercapacitor and BESS for peak shaving applications, In IEEE 15th Brazilian Power Electronics Conference and 5th IEEE Southern Power Electronics Conference (COBEP/SPEC), 1–6.

111. Song, Z., Hou, J., Hofmann, H., Li, J., & Ouyang, M., (2017) Sliding-mode and Lyapunov function-based control for battery/supercapacitor hybrid energy storage system used in electric vehicles, *Energy*, 122, 601–612.

112. Carignano, M.G., Costa-Castelló, R., Roda, V., Nigro, N.M., Junco, S., & Feroldi, D., (2017) Energy management strategy for fuel cell-supercapacitor hybrid vehicles based on prediction of energy demand, *Journal of Power Sources*, 360, 419–433.

113. Kafalis, K., & Karlis, A., (2017) Energy saving in elevators using flywheels or supercapacitors, *Recent Advances in Electrical & Electronic Engineering*, 10(1), 60–71.

114. Luri, S., Etxeberria-Otadui, I., Rujas, A., Bilbao, E., & Gonzalez, A., (2010) Design of a supercapacitor-based storage system for improved elevator applications, In IEEE Energy Conservation Congress and Exposition, 4534–4539.

115. Barrade, P., & Rufer, A., (2002) Supercapacitors as energy buffers: A solution for elevators and for electric busses supply, In Proceedings of the Power Conversion Conference, Osaka, 3, 1160–1165.

116. Partridge, J., & Abouelamaimen, D., (2019) The role of supercapacitors in regenerative braking systems, *Energies*, 12(14), 2683.

117. Burke, A., (2009) *CAPACITORS: Application, Chapter of Encyclopaedia of Electrochemical Power Sources*, Elsevier, 685–694.

118. Breeze, P., (2019) *Chapter 10 - Power System Energy Storage Technologies, Power Generation Technologies* (Third Edition), Newnes, 219–249.

119. Revankar, S.T., (2019) *Chapter Six - Chemical Energy Storage, Storage and Hybridization of Nuclear Energy*, Academic Press, 177–227.

120. Khan, M., Zeb, K., Sathishkumar, P., Ali, M., Uddin, W., Hussain, S., Ishfaq, M., Khan, I., Cho, H.G., & Kim, H.J., (2018) A novel supercapacitor/lithium-ion hybrid energy system with a fuzzy logic-controlled fast charging and intelligent energy management system, *Electronics*, 7(5), 63.

121. Perrotta, D., Ribeiro, B., Rossetti, R.J.F., & Afonso, J.L., (2012) On the potential of regenerative braking of electric buses as a function of their itinerary, *Procedia - Social and Behavioral Sciences*, 54, 1156–1167.

122. Wu, W., Bucknall, R.W.G., & Partridge, J., (2019) Development and evaluation of a degree of hybridisation identification strategy for a fuel cell supercapacitor hybrid bus, *Energies*, 12, 142.

123. Anh, A., & Duc, L., (2021) A regenerative braking energy recuperation from elevator operation in building by active rectifier, *International Journal of Power Electronics and Drive Systems (IJPEDS)*,12, 811.

124. Zhang, Y., Yan, Z., Yuan, F., Yao, J., & Ding, B., (2018) A novel reconstruction approach to elevator energy conservation based on a DC micro-grid in high-rise buildings, *Energies*, 12, 33.

125. Burke, A., (2010) Ultracapacitor technologies and application in hybrid and electric vehicles, *International Journal of Energy Research*, 34, 133–151.

126. Dubal, D.B., Chodankar, N.R., Kim, D.H., Gomez-Romero, P., (2018) Towards flexible solid-state supercapacitors for smart and wearable electronics, *Chemical Society Reviews*, 47, 2065–2129.

127. Wang, Y., Song, Y., & Xia, Y., (2016) Electrochemical capacitors: Mechanism, materials, systems, characterisation and applications, *Chemical Society Reviews*, 45, 5925–5950.
128. De, B., Banerjee, S., Verma, K.D., Pal, T., Manna, P.K., & Kar, K.K., (2020) Carbon nanofiber as electrode materials for supercapacitors, In Kar, K. (eds) *Handbook of Nanocomposite Supercapacitor Materials II, Springer Series in Materials Science*, Springer, 302.

9 Portable Electronics and Microsupercapacitors

Gopakumar G., Sujith K. V.,
Sarayu Jayadevan, and S. Anas

CONTENTS

9.1 INTRODUCTION

In the present era, we can't even imagine a world without modern portable electronic gadgets like smartphones, wireless devices, smartwatches, laptops, cameras, etc. Nowadays, the significance of flexible, wearable, and portable electronic devices is increasing in various sectors like smart electronics, consumer goods, sports, mobility, security and defence, medical and biomedical, green environment, clean energy, etc. [1–3]. Smart devices launched recently include not just computers or smartphones but every component of smart home technology.

Smart electronic devices generally require huge energy and need to be powered by efficient energy storage devices [2]. Conventional batteries help devices run for a long time on a single charge. The expeditious growth of smart and portable electronics demands flexible, lightweight, small, and wearable power sources. Capacitors are now becoming a key component of basic portable electronics as well as the most modern smart/hybrid electric vehicles by providing rapid delivery of energy despite their poor storage capacity [4]. Researchers are still working hard towards technologies to increase the storage ability of capacitors, strictly maintaining the green energy protocol. Supercapacitors (SCs) are found to be a better choice for addressing the energy issues of the portable electronic device industry. SCs bring

DOI: 10.1201/9781003258384-9

various benefits to electronic circuits. They help to (i) store energy to make charging/discharging faster, (ii) handle high voltage applications, (iii) filter out unwanted frequencies/fluctuations, and (iv) handle power loss situations [5].

Considering the advantages of SC systems including their stability, life span, and high power density without compromising the power density [6–9], the present chapter considers the various aspects of their application in portable electronics.

9.2 PORTABLE ELECTRONICS

SCs are emerging energy storage options with the potential to bridge the gap between rechargeable batteries and typical capacitors. SCs are the media of DC energy storage with short high-power storage of electrical charge. They are conventionally used to protect electronic devices during power cut situations as stabilizers of power supply analogous to UPS in computers. Owing to their ever-improving performance, SCs are regularly upgraded in their applications in energy systems of enhanced sustainability and lifetime. Based on their size, they are employed in various portable electronic devices such as personal digital assistants, satellite navigation devices, handheld devices, Wi-Fi appliances, etc. SCs work both as primary as well as a secondary sources of energy in systems with or without primary power sources. Remote controllers were initially powered by batteries which are being quickly replaced with SCs which have the capacity to induce rapid power outputs [1]. Also, SCs are used in digital cameras to produce flashes. Portable speakers working with wireless connectivity like Bluetooth are also powered by inbuilt SCs [2].

SCs can be used either as in combination with batteries or as a standalone power source wherever a sudden boost in current is needed [3]. Hybrid devices with battery/SC combinations are the best in supplying power for mobile electronic devices. This lightweight hybrid energy device can be fitted into small appliances like watches, mobile phones, sensors, headphones, other wireless communication equipment, etc. There have been many such small-scale combinations of batteries and capacitors in portable electronics [4]. Assemblages of SCs alone and in combination with batteries have been successfully employed to dispense energy backup to portable electronic devices like mobile phones, music players, tablets, remote controllers, laptops, screwdrivers, portable speakers, and portable palm-top scanners [5].

Earlier, SCs were used for memory backup for semiconductor memories, system boards, or microprocessors, and actuator applications in electronic devices including sensors. Later, they were replaced by microsupercapacitors (MSCs) with relatively small volumes and high electrochemical performances. tackle the need for compact and reliable power stabilizers in highly integrated portable devices, advanced wearable devices, and various sensors owing to their small size, high flexibility, low internal resistance, high charge and discharge rate, and light weight. MSCs are successfully applied in advanced sensors and related devices used for biological and medical purposes [6]. MSCs are used to energize various sensors as well as sensor integrations for different applications [7, 8].

Portable gas sensors coupled with MSCs can detect the presence of surrounding gases in real time and can be used for monitoring air quality measurements and human health. An MSC-gas sensor system was developed by Yun et al. by combining a graphene-based sensor with carbon nanotube (CNT)-based MSC [9]. Also, Song et al. introduced an assembly of piezoelectric sensors and MSC for detecting pressure in real time [10]. Moreover, MSCs have been successfully integrated with strain sensors, UV sensors, solar cells, etc., to extend their potential application [11].

Flexible and transparent SCs made of ultrathin graphene film are used in transparent electronic devices [5]. Various transparent electronic appliances like active-matrix LCD devices, active-matrix OLED devices, value-added glasses, and other cutting-edge products such as transparent smartphones are now using transparent SCs [12–14]. Also, transparent graphene film can be printed on substrates for portable and wearable electronics, with the potential for exciting future applications. Flexible SCs matching with the finish of electronic products may be the pointer towards the development of next-generation storage devices.

9.3 SUPERCAPACITORS IN WEARABLE ELECTRONICS

Wearable electronic technologies are intelligent devices attached to the surface of the body for detecting, analyzing, and transferring relevant information. Recently, carbon-based lightweight SCs are becoming new trends in the market of wearable electronics. These SCs can be integrated with advanced wearable technologies like smartwatches, biosensors, health monitoring devices, high-security clothing, and e-textiles [15]. Flexible SC devices have become popular with the rapid advancement of such technologies.

Wearable SCs are of different types based on their type of design and construction. They include:

• Coin or pouch SCs, used in smartwatches, etc.
• Printed SCs, printed in paper or fabric by technologies like screen or 3D printing
• Yarn-based SCs, used in e-textiles

The excellent lifetime and capacity to adapt to mechanical deformations exhibited by wearable SCs make them suitable for the area of wearable electronics [16, 17]. Wearable systems require energy autonomy which necessitates the presence of storage devices with enhanced energy density. Recently, high energy density flexible SCs based on graphene-graphite polyurethane composites have been developed successfully. They have found extensive applications in LEDs, wearable sensors, high torque motors, actuators of prosthetic limbs, etc. Paper-based electronic systems, consisting of piezoelectric generators and paper-based SCs, are found to be meeting the consumption requirements of novel electronic devices. They get activated on the movement of paper documents or the flipping of pages. These systems have been converted into a wireless interaction system that is capable of the management of documents and smart reading. This could support the Internet of Things (IoT) in communication and IT as well as the essentials of upgraded green electronics. A self-powered system with the capacity of simultaneous energy harvesting and sensing was developed from ultrathin piezoelectric hybrid nanogenerator (PNG) and paper-based SCs [18]. This exhibited a lot of potential applications for the IoT.

A polymer-based wearable SC has been developed on cloth that uses the wearer's sweat as the electrolyte by a sweat trap technology. The cloth in the system used flexible PEDOT:PSS polymers doped with some impurities and performed many cycles of charge and discharge without any deterioration. This has potential applications in robotics, IoT, and virtual reality [19].

Flexible SCs integrated with textiles have been successfully developed into mobile phone charging systems through the generation of piezoelectric charges. These SC-powered wearables can be charged by the movement of the wearer and can be ultimately used for charging mobile phones [20, 21]. Textile electronic articles are now used in sportswear, treadmill suits, military camouflage, work clothes, personal protective equipment (PPE), safety suits, and similar clothing. Printed SCs also find application in intelligent packaging as a power source [22]. Hence, we can assume that flexible SCs will expand their application to all future microelectronic products [23–25].

SCs played the role of energy source for smaller devices owing to their high power density, fast charge and discharge rates, along with long cycle lives. However, a fast-developing world demands electronic devices in their miniature forms for operational convenience and handling purposes. Therefore, much research is involved in designing and developing high-performance miniaturized SCs and their fabrication methods. Such SCs are known to be MSCs and the fabrication methods can be microfabrication. Some of such microfabrication methods are discussed below.

9.4 MICROSUPERCAPACITORS

9.4.1 FUNDAMENTALS OF MICROSUPERCAPACITORS

In general, a SC is composed of electrodes, an electrolyte, current collectors, and in many cases, a separator. The earlier stage of MSC fabrication was similar to thin-film capacitors with two film electrodes stacked with solid electrolytes in between giving a *sandwich-like* design whereas an *interdigitated planar* structure was developed later [26, 27].

9.4.1.1 Sandwich-Like Design

The concept was first demonstrated by Lim et al. in 2001 where a symmetric configuration with RuO_2 electrodes on both sides was sandwiched between lithium phosphorous oxynitride-based solid electrolyte. The system delivered a volumetric capacitance of ~380 $\mu F \cdot cm^{-3}$. Thin-film fabrication includes sputtering, chemical vapor deposition (CVD), layer-by-layer deposition (LBL), etc., and will be employed here to fabricate the electrodes [28]. However, the design has some practical drawbacks like the possibility of short circuits due to small distances between electrodes results in increased ion transport resistance in cells and leads to power loss.

9.4.1.2   In-Plane Interdigitated Design

The in-plane design is composed of finger electrodes in the same plane as that of the current collectors confiscated from each other by the narrow interspace with a gap width in the range of tens to hundreds of micrometres [29]. The electrolyte would be coated on top and also between the electrodes so as to guarantee ion transport which occurs along the basal plane of the electrodes. The design has more dominant advantages over the sandwich structure, and the most significant factor is that the peculiar structure provides a customizable performance matrix. Moreover, the diffusion length of the ions can easily be controlled by toggling the electrode width and the interspace which would alter the electrochemical series resistance (ESR). The in-plane configuration also prevents electrode short circuits [30, 31]. As organic binders and polymer separators are excluded in the configuration, planar interdigitated MSCs gain excellent mechanical and electrical properties. Therefore, the in-plane interdigital electrode design fabrication attained importance and dominance in the field of manufacture.

9.4.2   Fabrication Techniques for Interdigital Microsupercapacitors

Some general fabrication methods employed for the fabrication of planar MSCs are listed and discussed below [32].

   i)   Screen-printing method
  ii)   Ink-jet printing method
 iii)   Selective wetting-induced micropatterning fabrication method
  iv)   Microfluidic etching-assisted patterning fabrication method
   v)   Photolithography method
  vi)   Laser-irradiation-assisted patterning fabrication method

The *screen-printing method* is a simple and low-cost method used to print patterns on cloth and paper, where a woven mesh is used as an ink-blocking stencil to obtain desired images. It is considered to be the easiest and cheapest method for scalable production of MSCs [33, 34]. The mechanism of the process is quite simple, wherein ink pastes are penetrated through a patterned mask under a pressing force. The technique is widely used. For instance, Liu et al. fabricated a solid state MSC based on N-doped graphene and the electrode inks were screen printed into planar electrodes [35]. An areal capacitance of 3.4 mF·cm^{-2} with good rate capability was delivered. The quality of the process depends on the ink quality (viscosity and shear-thinning behaviour) and the resolution of the mask being used.

The *ink-jet printing method* is a simple patterning technique achieved with the help of a computer by propelling droplets of ink onto paper and plastics to achieve planar MSCs. Ink-jet printing does not need a mask or stencil and the ink droplets of micrometre size are ejected onto their appropriate position in the substrate using a micronozzle [36, 37]. The droplets are generally formed by thermal excitation. Piezoelectric excitation can also be used to generate the droplets. Li and co-workers reported a full-inkjet printing method in 2017 to fabricate MSCs based on graphene [38]. Ink-jet printing of graphene has been extensively studied and various kinds of pristine and hybrid graphene inks have been developed and successfully employed along with this technique.

The *selective wetting-induced micropatterning fabrication* method can be obtained for electrode-active materials dispersed in a hydrophilic solvent. Initially, in the fabrication process, an interdigitated pattern is made on a Si wafer. On to this interdigital finger-like microchannels cast polydimethylsiloxane (PDMS) [39, 40]. The number, width, and interspace between the interdigital microchannels can be altered to modify the performance of MSCs. The patterned side of the PDMS slab was exposed to oxygen plasma, making it hydrophilic. Peel off the micropatterned PDMS an aqueous multi-walled carbon nanotubes (MWNTs) suspension was carefully micropipetted into this interdigital finger-like microchannels. Then several drying-refilling-drying processes were continued to isolated microchannels of MWNT-patterned electrodes. The next process is drop-casting of a sufficient amount of PVA/H_3PO_4 solution onto the pattern. Once the mixture is solidified, peel off the PVA/H_3PO_4 film to obtain a flexible substrate with a solid electrolyte. Such devices exhibit excellent flexibility and long-term cycle stability with high power output and make a good candidate for high power on-chip energy storage applications [39].

In the *conventional photolithography method*, the patterns would be coated following the traditional photolithography process, whereby visible light is passed through a patterned mask into the resist coated substrate [41]. For instance, Au or Pt microelectrode arrays maintaining 50 μm width will be initially fabricated parallelly on Si substrate using photolithography and the wet-etching method. On these, conducting polymers such as polypyrrole (PPy) and poly-(3-phenylthiophene) (PPT) will be coated along with different electrolytes such as 0.1 M aqueous-based H_3PO_4 electrolytes and 0.5 M non-aqueous-based Et_4NBF_4/acetonitrile [42]. Cell performance and capacitance will be improved by altering the electrolytes and polymers. Microfluidic etching-assisted patterning is an easy soft lithography method, used to fabricate planar pseudo-MSCs. In one such approach, nanofibre-based MnO_2 film was prepared on a micro-Au-electrode collector deposited on a glass slide using the electrospinning technique [43]. The MnO_2 layer was then transferred to H_3PO_4/polyvinyl alcohol films acting as the electrolyte as well as the substrate. On to this network, a thin film of indium tin oxide (ITO) was evaporated using magnetic sputtering. Finally, MnO_2/ITO interdigital fingers on the solid electrolyte film were fabricated using the microfluidic etching method.

Laser-irradiation assisted patterning fabrication also achieves much more attention as an easy and cheaper alternative in this regard. For instance, Gao et al. reported a scalable fabrication of rGO-GO-rGO MSCs with in-plane and sandwich geometries based on CO_2 laser-patterning of free-standing hydrated graphene oxide (GO) films. Such monolithic MSCs by laser reduction and patterning of graphite oxide films is operationally convenient [26]. Using one-step and environmentally friendly ways, flexible energy storage devices can be fabricated using laser-assisted methods. The direct laser writing method has been widely applied to the fabrication of graphene-based, in-plane interdigitated MSCs, prepared on various substrates like free-standing GO paper (10 mm thick). Compared to conventional lithographic techniques, laser fabrication is faster, cost effective, and scalable. Another example for microfabrication is the fabrication of ultra high rate, all solid state, planar interdigital graphene-based MSCs manufactured by methane plasma-assisted reduction [44].

9.4.3 ELECTRODE MATERIALS

There is scope for an additional blend of different electrode materials with numerous planar MSCs with superb electrochemical execution. The feasibility and advantage of SCs are the materials used in the devices. For example, graphene and its derivatives are associated with some fine properties such as elasticity, porosity, transparency, intrinsic strength, conductivity, surface area along with light material nature. However, the major challenge associated with graphene-based ink is the restacking of individual sheets and much of the research is directed towards avoiding this issue. For instance, CNT fillers could effectively enlarge the inter layer spacing of graphene which would also improve the active surface area and energy profile [45]. Poly (3,4-ethylenedioxythiophene): poly (styrenesulphonic acid) (PEDOT:PSS) is another potential electrode material which would stabilize graphene, while also improving the electrochemical performance by acting as an active electrode component [46]. GO is another graphene-based ink for printing MSCs and is highly advantageous as the surface oxygen functional groups of GO would allow the formation of a well-dispersed ink in polar solvents like water, acetone, ethanol, and ethylene glycol [47]. One of the most recent trends in the field is the development of a unique class of 2D layered materials: MXenes, which are carbides or nitrides of transition metal oxides. They have the general formula $M_{n+1}X_nT_x$ where M is a transition metal (typically Ti, Mo, Tb, etc.), X is C or N, and T_x is a functional group (like OH, F etc.). Mxenes have several advantages. (1) They act as the active material as well as current collector owing to high conductivity. (2) The transition metal forms a reversible redox couple which imparts a pseudocapacitive storage mechanism, enhancing the storage capacity. (3) The hydrophilic groups on the surface of the layers avoid the need for any surfactant or additive to form a stable viscous and well dispersed ink [48].

9.5 SUMMARY AND OUTLOOK

There is an ever-increasing demand for novel and more efficient portable electronics and the market is growing rapidly. The chapter reviews the recent developments and trends in the field with particular emphasis on wearable electronics. In the context of the rapid development of miniature devices, a detailed discussion on the principles, fabrication, and material aspects of MSCs have also been included in the chapter. EDLCs as well as pseudocapacitive electrodes are being explored as potential candidates for MSC electrodes, whereas the major criterion is to make a well-dispersed ink in the case of planar MSCs. It is considered that configuring the device components into a 3D architecture could considerably improve the energy performance of the device; however, the developments in this regard are still in their infancy and efforts are to be focused in this direction. Further developments are to be aimed at integrating MSCs with micro-miniaturized devices, which in turn would require better materials and methods like leakage-free electrolytes and efficient patterning techniques. The increasing number of practical applications would surely encourage the researchers to address these challenges.

REFERENCES

1. Choi, C., et al (2020). Achieving high energy density and high power density with pseudocapacitive materials. *Nature Reviews Materials*, *5*, 5–19.
2. Kollimalla, S. K., Mishra, M. K., & Narasamma, N. L. (2014). Design and analysis of novel control strategy for battery and supercapacitor storage system. In *IEEE Transactions on Sustainable Energy*, *5*(4), 1137–1144.
3. Nguyen, C. M., Mays, J., Plesa, D., Rao, S., Nguyen, M., & Chiao, J. (2015). Wireless sensor nodes for environmental monitoring in Internet of Things. In IEEE MTT-S International Microwave Symposium, 1–4.
4. Khan, K., Tareen, A. K., Aslam, M., Mahmood, A., Khan, Q., Zhang, Y., Ouyang, Z., Guo, Z., & Zhang, H. (2020). Going green with batteries and supercapacitor: Two dimensional materials and their nanocomposites-based energy storage applications. *Progress in Solid State Chemistry*, *58*, 100254.
5. Bu, F., Zhou, W., Xu, Y., Du, Y., Guan, C., & Huang, W. (2020). Recent developments of advanced micro-supercapacitors: Design, fabrication and applications. *Npj Flexible Electronics*,*4*, 31.
6. Muzaffar, A., Ahamed, M. B., Deshmukh, K., & Thirumalai, J. (2019). A review on recent advances in hybrid supercapacitors: Design, fabrication and applications. *Renewable and Sustainable Energy Reviews*, *101*(C), 123–145.
7. Zhao, C., Liu, Y., Beirne, S., Razal, J. & Chen, J (2018). Recent development of fabricating flexible micro-supercapacitors for wearable devices. *Advanced Materials Technologies*, *3*, 1800028.
8. Yun, J. et al (2017). A patterned graphene/ZnO UV sensor driven by integrated asymmetric micro-supercapacitors on a liquid metal patterned foldable paper. *Advanced Functional Materials*, *27*, 1700135.
9. Yun, J. et al (2016). Stretchable patterned graphene gas sensor driven by integrated micro-supercapacitor array. *Nano Energy*, *19*, 401–414.
10. Song, Y. et al (2018). All-in-one piezoresistive-sensing patch integrated with micro-supercapacitor. *Nano Energy*, *53*, 189–197.
11. Kaidarova, A. et al (2019). Wearable multifunctional printed graphene sensors. *NPJ Flexible Electronics*, *3*, 15.
12. Chamoli, P., Das, M. K., & Kar, K. K. (2018). Urea-assisted low temperature green synthesis of graphene nanosheets for transparent conducting film. *Journal of Physics and Chemistry of Solids*, *113*, 17–25.
13. Guan, X., Pan, L., & Fan, Z. (2021). Flexible, transparent and highly conductive polymer film electrodes for all-solid-state transparent supercapacitor applications. *Membranes*, *11*(10), 788.
14. Khan, M. I., Bibi, F., Hassan, M. M., Muhammad, N., Tariq, M., & Rahim, A. (2021). Chap 10- Inorganic electrodes for flexible supercapacitor. In: *Flexible Supercapacitor Nanoarchitectonics*, Scrivener Publishing, 263–275.
15. Wang, Y., Feng, C., Yang, J., Zhou, D., & Liu, W. (2020). Static response of functionally graded graphene platelet–reinforced composite plate with dielectric property. *Journal of Intelligent Material Systems and Structures*, *31*(19), 2211–2228.

16. Philip, N. Y., Rodrigues, J. J. P. C., Wang, H., Fong, S. J., & Chen, J (2021). Internet of things for in-home health monitoring systems: Current advances, challenges and future directions. *IEEE Journal on Selected Areas in Communications*, *39*, 300.

17. Choi, C., Myeong Lee, J., Kim, S. H., Kim, S. J., Di, J., & Baughman, R. H. (2016). Twistable and stretchable sandwich structured fiber for wearable sensors and supercapacitors. *Nano Letters*, *16*(12), 7677–7684.

18. Manjakkal, L., Navaraj, W. T., Núñez, C. G., & Dahiya, R. (2019). Graphene- graphite polyurethane composite based high- energy density flexible supercapacitors. *Advanced Science*, *6*(7), 1802251.

19. He, X., Zi, Y., Hua, Y., Zhang, S. L., Wang, J., Ding, W., Zou, H., Zhang, W., Lu, C., & Wang, Z. L. (2017). An ultrathin paper-based self-powered system for portable electronics and wireless human-machine interaction. *Nano Energy*, *39*, 328–336.

20. Manjakkal, L., Pullanchiyodan, A., Yogeswaran, N., Seyed Hosseini, E., & Dahiya, R (2020). A wearable supercapacitor based on conductive PEDOT:PSS-coated cloth and a sweat electrolyte. *Advanced Materials*, *32*(24), 1907254.

21. Pu, X., & Wang, Z. L. (2020). Self-charging power system for distributed energy: beyond the energy storage unit. *Chemical Science*, *12*(1), 34–49.

22. Jost, K., Stenger, D., Perez, C. R., McDonough, J. K., Lian, K., Gogotsi, Y., & Dion, G. (2013). Knitted and screen printed carbon-fiber supercapacitors for applications in wearable electronics. *Energy & Environmental Science*, *6*, 2698–2705.

23. Tahalyani, J., Akhtar, J., Cherusseri, J., & Kar, K. K. (2020). Chap 1- Characteristics of capacitor: Fundamental aspects. In: *Handbook of Nanocomposite Supercapacitor Materials I*, Springer, 1–51.

24. Huang, S., Zhu, X., Sarkar, S., & Zhao, Y. (2019). Challenges and opportunities for supercapacitors. *APL Materials*, *7*, 100901.

25. Yang, P., & Mai, W. (2014). Flexible solid-state electrochemical supercapacitors. *Nano Energy*, *8*, 274–290.

26. Gao, W., Singh, N., Song, L., Liu, Z., Reddy, A. L. M., Ci, L, & Ajayan, P. M. (2011). Direct laser writing of micro-supercapacitors on hydrated graphite oxide films. *Nature Nanotechnology*, *6*(8), 496–500.

27. Su, Z., Yang, C., Xie, B., Lin, Z., Zhang, Z., Liu, J, & Wong, C. P. (2014). Scalable fabrication of MnO 2 nanostructure deposited on free-standing Ni nanocone arrays for ultrathin, flexible, high-performance micro-supercapacitor. *Energy & Environmental Science*, *7*(8), 2652–2659.

28. Lim, J. H., Choi, D. J., Kim, H. K., Cho, W. I., & Yoon, Y. S. (2001). Thin film supercapacitors using a sputtered RuO2 electrode. *Journal of The Electrochemical Society*, *148*(3), A275.

29. Liu, N., & Gao, Y. (2017). Recent progress in micro-supercapacitors with in-plane interdigital electrode architecture. *Small*, *13*(45), 1701989.

30. Mendoza-Sánchez, B., & Gogotsi, Y. (2016). Synthesis of two-dimensional materials for capacitive energy storage. *Advanced Materials*, *28*(29), 6104–6135.

31. Xiong, G., Meng, C., Reifenberger, R. G., Irazoqui, P. P., & Fisher, T. S. (2014). A review of graphene-based electrochemical microsupercapacitors. *Electroanalysis*, *26*(1), 30–51.

32. Hu, H., Pei, Z., & Ye, C. (2015). Recent advances in designing and fabrication of planar micro-supercapacitors for on-chip energy storage. *Energy Storage Materials*, *1*, 82–102. DOI: 10.1016/j.ensm.2015.08.005.

33. Hyun, W. J., Secor, E. B., Hersam, M. C., Frisbie, C. D., & Francis, L. F. (2015). High-resolution patterning of graphene by screen printing with a silicon stencil for highly flexible printed electronics. *Advanced Materials*, *27*(1), 109–115.

34. Wang, Y., Shi, Y., Zhao, C. X., Wong, J. I., Sun, X. W., & Yang, H. Y. (2014). Printed all-solid flexible microsupercapacitors: Towards the general route for high energy storage devices. *Nanotechnology*, *25*(9), 094010.

35. Liu, S., Xie, J., Li, H., Wang, Y., Yang, H. Y., & Zhu, T, X. (2014). Nitrogen-doped reduced graphene oxide for high-performance flexible all-solid-state micro-supercapacitors. *Journal of Materials Chemistry A*, *2*(42), 18125–18131.

36. Singh, M., Haverinen, H. M., Dhagat, P., & Jabbour, G. E. (2010). Inkjet printing: Process and its applications. *Advanced Materials*, *22*(6), 673–685.

37. Calvert, P. (2001). Inkjet printing for materials and devices. *Chemistry of Materials*, *13*(10), 3299–3305.

38. Li, J., Sollami Delekta, S., Zhang, P., Yang, S., Lohe, M. R., Zhuang, X., & Ostling, M. (2017). Scalable fabrication and integration of graphene microsupercapacitors through full inkjet printing. *ACS Nano*, *11*(8), 8249–8256.

39. Hu, H., Pei, Z., & Ye, C. (2015). Recent advances in designing and fabrication of planar micro-supercapacitors for on-chip energy storage. *Energy Storage Materials*, *1*, 82–102.

40. Wang, F., Wu, X., Yuan, X., Liu, Z., Zhang, Y., Fu, L., & Huang, W. (2017). Latest advances in supercapacitors: from new electrode materials to novel device designs. *Chemical Society Reviews*, *46*(22), 6816–6854.

41. Kurra, N., Jiang, Q., & Alshareef, H. N. (2015). A general strategy for the fabrication of high performance microsupercapacitors. *Nano Energy*, *16*, 1–9.

42. Beidaghi, M., & Gogotsi, Y. (2014). Capacitive energy storage in micro-scale devices: Recent advances in design and fabrication of micro-supercapacitors. *Energy & Environmental Science*, *7*(3), 867–884.

43. Xue, M., Xie, Z., Zhang, L., Ma, X., Wu, X., Guo, Y., & Cao, T. (2011). Microfluidic etching for fabrication of flexible and all-solid-state micro supercapacitor based on MnO 2 nanoparticles. *Nanoscale*, *3*(7), 2703–2708.

44. Paquin, F., Rivnay, J., Salleo, A., Stingelin, N., & Silva, C. (2013). Multi-phase semicrystalline microstructures drive exciton dissociation in neat plastic semiconductors. *arXiv preprint arXiv:1310.8002*.

45. Yang, S. Y., Chang, K. H., Tien, H. W., Lee, Y. F., Li, S. M., Wang, Y. S., & Hu, C. C. (2011). Design and tailoring of a hierarchical graphene-carbon nanotube architecture for supercapacitors. *Journal of Materials Chemistry*, *21*(7), 2374–2380.

46. Liu, Z., Wu, Z. S., Yang, S., Dong, R., Feng, X., & Müllen, K. (2016). Ultraflexible in-plane micro-supercapacitors by direct printing of solution-processable electrochemically exfoliated graphene. *Advanced Materials*, *28*(11), 2217–2222.

47. Zhu, Y., Murali, S., Cai, W., Li, X., Suk, J. W., Potts, J. R., & Ruoff, R. S. (2010). Graphene and graphene oxide: Synthesis, properties, and applications. *Advanced Materials*, *22*(35), 3906–3924.

48. Peng, Y. Y., Akuzum, B., Kurra, N., Zhao, M. Q., Alhabeb, M., Anasori, B., ... & Gogotsi, Y. (2016). All-MXene (2D titanium carbide) solid-state microsupercapacitors for on-chip energy storage. *Energy & Environmental Science*, *9*(9), 2847–2854.

10 Electric and Hybrid Electric Vehicle

Sruthi Maruthiyottu Veettil,
Sindhu Thalappan Manikkoth, Anjali Paravannoor,
and Baiju Kizhakkekilikoodayil Vijayan

CONTENTS

10.1 INTRODUCTION

We are going to exit the fossil fuel era. It is inevitable.

Elon Musk

Expeditious urbanization caused a tremendous increase in the population of cities, which crucially increased mobility and vehicles on the roads. An increase in the number of vehicles caused increased CO_2 emissions. An ordinary vehicle evaporates heat during utilization of 85% of fuel energy in terms of CO_2, one of the greenhouse gases that causes global warming. Eventually climate changes cause environmental problems [1]. In order to solve this problem, we need advanced technology in the transport sector. So electric vehicles (EVs) are

DOI: 10.1201/9781003258384-10

the best contemporary solution due to zero emission. EVs are vehicles that are either partially or completely powered by electric power. In order to decrease the carbon footprint and pollution there is no doubt EVs are the future of the world for sustainable development [2].

You might wonder what is the importance of supercapacitors (SCs) in EVs? Why are we discussing EVs? We know the common energy storage system (ESS) in vehicles is batteries. But batteries are not enough for today's energy demands because of low efficiency. So, SCs are the best alternative to solve these problems. SCs are used in utilization, where we need to store or release a large amount of energy in a short time. Due to high efficiency, high charge acceptance, as well as easy charge/discharge properties, SCs are the future [3]. Nowadays SCs are used mainly in EVs, fuel cell vehicles like trains and cars. Another area of application is in electronic devices such as uninterruptible power supply (UPS) and volatile memory backups in computers. The third area of application is in energy harvesting systems like, wind turbines and solar panels. Recently Tesla bought Maxwell Technologies which is the one of the best manufactures of SCs in the world. Both batteries and SCs can be used for charging EVs, but batteries will take average time of 12 hours, and SCs take 30 minutes to charge fully. Due to their large charge and discharge property, they can accelerate a vehicle faster [4].

10.2   MODERN ELECTRIC VEHICLES

EVs are the vehicles that make use of one or more electric motors for propulsion. Nowadays EVs are not restricted to roads, they are included in electric railways, electric spacecraft too. The history of EVs starts from more than 100 years ago. In 1828, Anyons designed an early type of electric motor, and created a small car. After that, officially the first electric car was made in 1884. In history the EV was awarded the unique significance of becoming the first manned vehicle to drive on the Moon; it was the Lunar Roving Vehicle, which was first developed during the Apollo-15 mission. Even though mass production of EVs started with the introduction of the Toyota Prius in Japan in 1997, another factor which helped EVs on their long journey was Tesla Motors, a small Silicon Valley startup started producing a luxury electric sports car that could go more than 200 miles on a single charge [5, 6].

10.2.1   Major Types of Electric Vehicles

One of the major advantages of EVs are their better energetic performance, EVs use less energy than conventional vehicles which directly depend upon the type of EV. Based on the primary energy used for the production of electricity, five types of EV are identified and are given in Scheme 10.1 [7].

10.2.1.1   Hybrid Electric Vehicle (HEV)

HEVs are vehicles that make use of more than one method of propulsion, combining a fuel engine along with an electric motor. One of the main advantages of hybrid

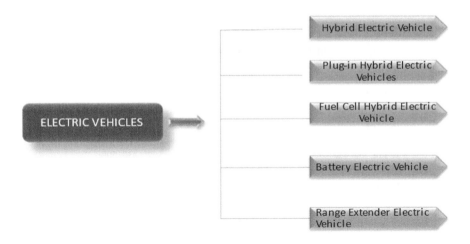

SCHEME 10.1 Classification of electric vehicle based on primary energy used for energy production.

vehicles are they use less fuel and emit fewer greenhouse gases than conventional vehicles. HEVs usually have an engine with a fuel tank, a motor, and a battery. The electric system gives higher acceleration performance at low speed. The flow of power in and internal combustion engine (ICE) is unidirectional, from engine to wheels, but in HEVs it can be bidirectional, this means from motor to wheel and wheel to battery. In order to find how much power is shared, the *hybridization factor* *(HF)* is introduced, which is defined as follows:

$$HF = \frac{\text{Sum of Power of Electric Motors}}{\text{Sum of Motor Power} + \text{Engine Power}}$$

On the basis of degree of hybridization, HEVs can again be classified into three [8]:

a) Micro hybrid vehicle (μHV). It is the least electrified type, usually used for continuous idle-stop or stop-start mode operations. It is a normal ICE vehicle with huge starter motor (about 3–5 kW at 12 V) to aid the starting of the ICE. Even though the motor cannot move the vehicle, it can assist components like air conditioning and power steering [9].

b) Mild hybrid vehicle (MHV). Like a μHV motor (about 7–15 kW at 60–200 V) alone cannot move the vehicle, but can support the starting of the engine, regenerative braking, and also give sufficient torque when high power is needed during acceleration.

c) Full hybrid vehicle (FHV). It can run only in electric mode which needs a large capacity motor (about 30–50 kW at 200–600 V). Energy savings of this type of vehicle are about 50%.

A comparison of different types of HEVs is given in Table 10.1.

TABLE 10.1

Comparison of Different Types of Hybrid Electric Vehicles

	Micro Hybrid	Mild Hybrid	Full Hybrid
Motor Power	3–5 kW	7–15 kW	30 kW
Motor Voltage	12 V	60–200 V	200–600 V
Hybridization	1–10%	10–30%	40%
IC Engine	Conventional	Downsized	Downsized
Energy Savings	5–10%	20–30%	30–50%
Cost (relative)	Low	Medium	High
Examples	Mercedes Smart	Honda Insight	Toyota Prius

10.2.1.2 Plug-In Hybrid Electric Vehicle (PHEV)

Plug-in hybrid electric vehicles (PHEVs) have almost same powertrain as HEVs. The main difference is that PHEVs usually use batteries to power an electric motor and use another fuel; gasoline, to power an ICE. The batteries can be charged using a charging station, by the ICE, or through the regenerative breaking. PHEVs run on electric power until the battery is exhausted, and then the vehicle instantly switches to use the ICE. Compared to standard HEVs, all electric ranges are greater for PHEVs. Due to high fuel efficiency, they have a more extended driving range than EVs, less running cost, etc. PHEVs are environmentally friendly. The Ford CMax Energi, Toyota Prius, and Chevy Volt are examples of this category [3, 9, 10].

10.2.1.3 Fuel Cell Hybrid Electric Vehicle (FCHEV)

Fuel cell hybrid electric vehicles (FCHEVs) have a series hybrid arrangement in which a fuel cell energy transfiguration system and battery or SC are the ESS to provide acceleration power. FCHEVs generate electricity using a fuel cell, which is powered by hydrogen, instead of producing electricity only from the battery. At present FCHEV technology is very expensive and too early. The Honda Clarity and Hyundai Santa Fe FCHEVs are on the market so far.

10.2.1.4 Battery Electric Vehicle (BEV)

Battery electric vehicles (BEVs) (or all-electric vehicles) are accelerated by electric motors by using energy stored in batteries. BEVs and ICE vehicles are almost similar, but in the case of the recharging of BEVs, they must systematically be plugged into an external source. Different models such as Nissan Leaf, Renault Twizy, Citroen C-Zero, and Tesla, are available on the market [6].

10.2.1.5 Range Extender Electric Vehicle (REXEV)

Range extender electric vehicles (REXEVs) are vehicles which are propelled mainly by electric propulsion but use an ICE to assist the battery and extend the range. The range extender is an independent power unit put onto pure electric drive vehicles to increase their operational range. There are only few examples, like Chevrolet Volt, Opel Ampera, etc. [6].

10.3 STORAGE SYSTEMS FOR ELECTRIC VEHICLE APPLICATIONS

In order to reduce CO_2 emissions and increase the use of renewable energy resources, we need a highly efficient ESS. An ideal ESS should provide a continuous and flexible energy supply to support and intensify energy as a result of crowding and interruption of the transmission line for extra energy demands. In addition, it could guarantee service to consumers during power outages due to natural disasters. In the case of EVs, they are put off use of fossil fuels and try to reduce CO_2. So, for EVs high-potential ESSs are necessary. The most common choices are ultracapacitors and batteries. Batteries have high energy density, whereas ultracapacitors have high power density. EVs need a combination of these two properties [11, 12].

ESSs can be classified into chemical, electro-chemical, mechanical, electrical, thermal, and hybrid. According to formation and compositional materials these systems can again be divided. Here we will discuss fuel cell and SC storage systems. A simple diagrammatic representation in shown in Scheme 10.2.

10.3.1 FUEL CELLS

Chemical conversion systems are those which transform the chemical energy present in a substance to energy and other components through chemical

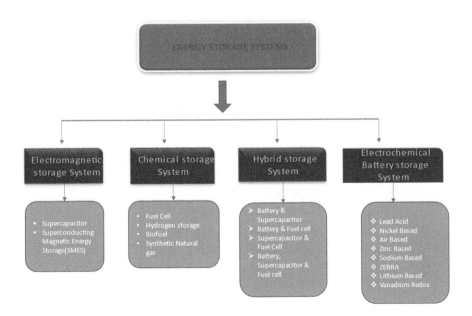

SCHEME 10.2   Diagrammatic representation of different types of energy storage system.

reactions. Fuel cells are electrochemical devices, without any intermediate step they can directly convert the chemical energy to electrical energy. As electrochemical conversion is not involved any other mechanical movement for the process leads to higher efficiency and a longer lifetime. Compared with an ordinary process, where the fuel's chemical energy is first converted into heat then into other useful forms, in the case of fuel cells, chemical energy is converted directly into electricity, so the fuel cell energy transformation process is not limited by the Carnot efficiency. However, the fuel cells also produce heat energy, but can be used for other purposes.

Fuel cells consist of two electrodes: cathode and anode, separated by an electrolyte. The major role of electrolytes is to enhance the transportation of cations and anions. The fuel cell operating temperature depends upon the type of electrolyte used. Reactions in different electrodes are as follows:

At anode:

$$2H_2 \Rightarrow 4H^+ + 4e^-$$

At cathode:

$$O_2 + 4e^- + 4H^+ \Rightarrow 2H_2O$$

Overall reaction:

$$2H_2 + O_2 \Rightarrow 2H_2O$$

With the evolution of different types of materials, different types of fuel cell can be constructed for various applications. The naming of the fuel cell is mostly based on the type of electrolyte used. Based on the electrolyte, fuel cells can be divided into five types [13], they are:

1. Alkaline Fuel Cell (AFC)
2. Polymer Electrolyte Fuel Cell (PEFC)
3. Phosphoric Acid Fuel Cell (PAFC)
4. Molten Carbonate Fuel Cell (MCFC)
5. Solid Oxide Fuel Cell (SOFC)

Among these types, a PAFC uses a medium temperature fuel cell (FC). And due to high efficiency, ease of design, and low emissions, these are very useful in transportation. SOFCs and MCFCs usually perform at high temperatures, about 600–1000°C. Because of this characteristic, these are widely used for large-scale energy storage and generation and grid applications. SOFCs have better stability as well as high fuel efficiency compared with others. That's why SOFCs are mostly used in EVs as a potential auxiliary power source.

10.3.2   Hybrid Storage System (HSS)

Based on characteristics and the diverse applications of different ESSs, its possibility for hybridization should researched. Some of the features are response time and efficiency balancing, life cycle, energy density, and power. So HSSs consist of two ESSs, like fast response ESSs and low response ESSs, high cost ESSs and low cost ESSs, etc. We can classify the HSS into battery and battery hybrid, battery and ultracapacitor (UC) hybrid, FC and battery hybrid, battery and superconducting magnetic energy storage (SMES) hybrid, battery and flywheel hybrid, FC and UC hybrid, etc. The most common one is battery and SC hybrid combination. HSSs have lots of benefits over other ESSs, like increase in system efficiency, longer lifetime, cost reduction, large storage capacities. Normally in EVs, battery-FC and battery-SC hybrids are used for better performance. For powering EVs, FC, battery, or UC can be used. Based on the EV properties and requirement, nickel-based batteries, silver batteries, sodium-sulfur batteries, lithium-ion batteries, lead-acid batteries, and SCs are used [11, 14].

10.3.3   Hybrid Supercapacitors for EV Applications

The conceptualization of hybrid SCs evolved to improvise the energy density of SCs to a range of 20–30 Wh kg^{-1} thus building an ESS with both energy and power efficiency [15]. The hybrid SCs are made by coupling an EDLC and pseudocapacitors by retaining or improving the excellent characteristics and eclipsing the limitations of their individual components [16, 17].

The energy storage mechanism of hybrid SCs is the combination of charge storage in the double layer of active material from the EDLC portion and fast repetitive faradaic redox reactions between the electrode and electrolyte from the pseudocapacitive portion. Such a hybrid capacitor system exhibits high energy and power density, wide operating voltage windows, and thus higher capacitance compared to their individual components, resulting in ameliorated performance in energy storage applications [18, 19]. A schematic illustration of a hybrid SC with a carbon electrode and lithium-ion insertion electrode is shown in Figure 10.1.

Like conventional SCs, hybrid assembled capacitors can also be symmetric or asymmetric depending upon the electrode configuration employed in the device fabrication [21–24]. Better results are achieved with asymmetric SCs where two different electrodes with different active materials have been used [25, 26]. The higher specific capacitance value with a wide potential window is achieved with the hybrid configuration, unlike conventional EDLCs, whereas excellent cyclic stability with a greater extent of affordability is achieved with the hybrid SC which is accounted as the limiting factor of pseudocapacitors [27].

The wide operating temperature and potential range and long charge/discharge cycles together with enhanced capacitance and exceptional energy storage ability of hybrid SCs make them superior candidates in EV applications over batteries and FCs [4]. The hybrid SCs can easily produce a high current in a short time, generating an instantaneous power pulse which is the essential requirement for charging in EVs

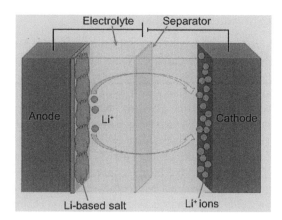

FIGURE 10.1 Schematic illustration of a hybrid supercapacitor [20]. R*eproduced from Chen, X., Paul, R., & Dai, L. (2017). Carbon-based supercapacitors for efficient energy storage. National Science Review, 4(3), 453–489 under the terms of the Creative Commons CC BY license [20].*

or HEVs. Toyota and Mazda manufacture cars based on efficient hybrid SCs. The Toyotas Yaris and PSA Peugeot Citroëns are based on these hybrid SCs for starting and stopping and also for energy bursts which ensure fuel saving together with faster primary acceleration. Large EVs also use hybrid supercapacitors (HSC) with a rapid charging facility. One of the giants of the EV industry, Aowei Technology Co., Ltd (Shanghai, China), introduced electric trolley buses exclusively made up of hybrid SCswith a hybrid configuration of $Ni(OH)_2$-AC offered a distance range of 7.9 km with a maximum speed of 44.8 km/hr and a charging time of 90 sec [28]. The tram car introduced by CSR Co. Ltd (Puzhen, Nanjing city; China) with EDLC configuration provides a 3–5 km distance range with a charging time of 30 sec.

Regenerative braking technology using SCs is another important strategy towards the improved performance of EVs. In this technique, the braking energy is captured by stopping the vehicle. The technology was well used by trams in Switzerland, France, and Belgium by installing one-ton hybrid SC assembly and combining 48 SCs respectively, which exhibited a charging time of 20 sec. The regenerative braking mechanism needs a rapid energy storage device with a high cycle rate and thus hybrid SCs are ideal candidates for the application. The braking energy obtained is utilized instead of burning diesel engine and results in speeding up of the EVs with energy saving of up to 30%. Jung et al. introduced a SC and explored their performances in a 42-V automotive electrical system [29]. A 6-kW power-boosting/regenerative braking was achieved with the SC modules. Public transportation such as buses and trucks utilize the SC for acceleration purposes by reducing the heat energy wastage during the stop though the engine is running. This helps to minimize fuel combustion and thus CO_2 emission. The Sinautec company introduced electric buses with charging stations at bus stops, which is more convenient for the customers. Furthermore, the Airbus 380 jumbo jet evacuation slide operation and emergency door operation confirms the potential use of hybrid SCs in EVs. The light metro introduced by CSR Zhuzhou Electric Locomotive Corporation of China uses hybrid SCs mounted on the roof for its operation [19]. Many researchers are investigating various materials and configurations for efficient hybrid SCs for EV applications. Dong et. al fabricated safe, high-rate, and ultralong life zinc-ion hybrid SCs that can be used for EV applications [30]. Nawa technologies (France) claims that the hybrid capacitors will be capable of doubling the range of existing EVs to 1000 kilometers with a battery that can recharge to 80% in five minutes. They also declared that the system is able to last for over 1,000,000 cycles compared with lithium-ion batteries which last for only between 300 and 500. Research is going on to fabricate efficient hybrid capacitors for EVs. Thus there is an ever-increasing scope for electrifying vehicles.

10.4   THE MODELING OF SUPERCAPACITORS IN EVS

SCs have high energy density, about 1000 times greater than a conventional capacitor. The structure of a basic SC cell is cylindrical, even though due to the achievements of technology commercial pouch SCs are also available on the market. We already discussed the basic principles and working of SCs in previous chapters. For high power automobiles, we need to connect SCs in parallel with the battery. In order to improve the life cycle and performance of SCs in EVs systems, the development and simulation of models are very important [4, 31, 32].

10.4.1   ELECTRIC MODELS

Different SCs are developed for different applications. A model has been proposed by Faranda et al. This model contains three branches. First branch is R_0 and represents the fast response of SCs in terms of seconds. The second branch consists of capacitor and resistance. The second branch analyses the long-term nature of SCs in terms of minutes. The simple equivalent circuit model is shown in Figure 10.2(a). But there is a

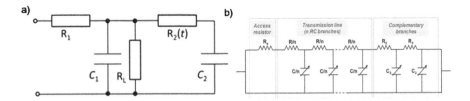

FIGURE 10.2   (a) Equivalent circuit model [34]. *Reproduced from Sedlakova, V., Sikula, J., Majzner, J., Sedlak, P., Kuparowitz, T., Buergler, B., & Vasina, P. (2015). Supercapacitor equivalent electrical circuit model based on charge redistribution by diffusion. Journal of Power Sources, 286, 58–65 with permission from Elsevier [34].* (b) Multi-branch model [35]. *Reproduced from Logerais, P. O., Camara, M. A., Riou, O., Djellad, A., Omeiri, A., Delaleux, F., & Durastanti, J. F. (2015). Modeling of a supercapacitor with a multibranch circuit. International Journal of Hydrogen Energy, 40(39), 13725–13736 with permission from Elsevier [35].*

considerable (about 10%) error for this model. [32, 33]. In order to minimize drawbacks, we can use a multi-level branch model, which is shown in Figure 10.2(b). Thus, second type of electric model is for evaluating SC behavior. Name of this model is the Thevenin battery model, which has higher accuracy than previous ones due to non-linear behavior. Another model is also proposed based on electrochemical characterization of SCs on electrode and electrolyte level. One thing we should keep in mind is this model needs lot of test procedures to determine model parameters, which can only be carried out by a chemist. So, we can't use this model for vehicular applications.

In the case of the first two models, parameters can easily be found out by the electrical method. But due to complexity and simulation time, it is very difficult to use in HEV applications. So, there is an interesting alternative model in real application, called the RRC model. Where C_0 is a constant, ESR is equivalent series resistance, $Ck = k*V$ varies with SC voltage. This model is very suitable for application, where energy storage in the capacitor is of primary importance. A circuit diagram of the RRC model is shown in Scheme 10.3.

10.4.2 THERMAL MODELING

Based on thermal-electric analogy, the thermal model determines the SC temperature inside and at the surface. This developed model can easily be implemented in different simulation programs. This model helps to study the heat management in SCs. This model consists of number of series or parallel arrangements of cells. The aim of this model is to locate and calculate the maximum temperature in order to size the cooling system [3, 36]. A simple representation of the thermal-electric model of SC is shown in Figure 10.3.

With the thermal model, we can find the temperature on the external surface on SCs depending on the power, the ambient temperature, and heat transfer coefficient. The total power dissipated in SC, can be calculated by using the equation:

$$P(t) = ESR \times I(t)$$

Where ESR is the equivalent series resistance of the SC and $I(t)$ is the Root Mean Square (RMS) current passing through the SC. In circuit model R_{conv} represents heat conversion between the SC surface and atmosphere air.

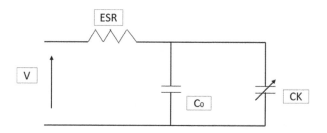

SCHEME 10.3   Circuit diagram of RRC model [32]

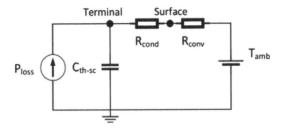

FIGURE 10.3 Circuit diagram of thermal-electric model of a supercapacitor [32]. *Supercapacitor reproduced from Župan, I., Šunde, V., Ban, Ž., &Krušelj, D. (2021). Algorithm with temperature-dependent maximum charging current of a supercapacitor module in a tram regenerative braking system. Journal of Energy Storage, 36, 102378 with permission from Elsevier [32].*

10.5 SUMMARY

This chapter discusses the major EVs, their energy storage mechanisms, and modeling of novel configurations. The integration of onboard ESSs with enhanced SC devices can lead to significant advancements in EVs with regard to enhanced performance, environmental impact, energy economy, and extended battery life. Diverse device architectures are established to resolve the economic and technical drawbacks to achieving improved results. However, for the increased demand for EVs, limitations such as high cost, low driving range, and lack of charging stations must be overcome. SCs already aid in the fast charging of electric buses with their drives from stop to stop. Such charging systems are not made on a commercial scale; nevertheless, people buy the vehicles with a belief that the development of large-scale SC chargers will likely happen. So, we can hope that SC charging points and wireless charging pads will be available at stations, like a petrol pump able to fill up a car in minutes. The idea can be developed to get a wireless zip line charging facility on the road which ensures charging during driving. The core objective of this chapter is to explore the future market scenarios of various kinds of EVs with an economic balance with conventional cars. Additionally, the environmental prospects and improvement in driving range are very important as they are key factors in the promotion of EVs.

REFERENCES

1. Teixeira, A. C. R., & Sodré, J. R. (2018). Impacts of replacement of engine powered vehicles by electric vehicles on energy consumption and CO_2 emissions. *Transportation Research Part D: Transport and Environment, 59,* 375–384.
2. Perujo, A., & Ciuffo, B. (2010). The introduction of electric vehicles in the private fleet: Potential impact on the electric supply system and on the environment. A case study for the Province of Milan, Italy. *Energy Policy, 38*(8), 4549–4561.
3. Hannan, M. A., Azidin, F. A., & Mohamed, A. (2014). Hybrid electric vehicles and their challenges: A review. *Renewable and Sustainable Energy Reviews, 29,* 135–150.
4. Kouchachvili, L., Yaïci, W., & Entchev, E. (2018). Hybrid battery/supercapacitor energy storage system for the electric vehicles. *Journal of Power Sources, 374,* 237–248.

5. Horn, M., MacLeod, J., Liu, M., Webb, J., & Motta, N. (2019). Supercapacitors: A new source of power for electric cars? *Economic Analysis and Policy*, *61*, 93–103.

6. Ajanovic, A. (2015). The future of electric vehicles: Prospects and impediments: The future of electric vehicles. *Wiley Interdisciplinary Reviews: Energy and Environment*, *4*(6), 521–536.

7. Ehsani, M. (2018). *Modern Electric, Hybrid Electric, and Fuel Cell Vehicles* (Third edition.). Taylor & Francis, CRC Press.

8. *Electric Vehicles: Prospects and Challenges*. (2017). Elsevier.

9. Vidyanandan, K. V. (2018). Overview of electric and hybrid vehicles. *Energy Scan*, *3*, 7–14.

10. Burke, A. (2010). Ultracapacitor technologies and application in hybrid and electric vehicles. *International Journal of Energy Research*, *34*(2), 133–151.

11. Habib, A. K. M. A., Hasan, M. K., Mahmud, M., Motakabber, S. M. A., Ibrahimya, M. I., & Islam, S. (2021). A review: Energy storage system and balancing circuits for electric vehicle application. *IET Power Electronics*, *14*(1), 1–13.

12. A novel design of hybrid energy storage system for electric vehicles. (2018). *Chinese Journal of Electrical Engineering*, *4*(1), 45–51.

13. Biradar, S. K., Patil, R. A., & Ullegaddi, M. (1998). Energy storage system in electric vehicle. In Power Quality '98 (pp. 247–255). Presented at the 1998 Power Quality Conference, Hyderabad, India: IEEE.

14. Hannan, M. A., Hoque, M. M., Mohamed, A., & Ayob, A. (2017). Review of energy storage systems for electric vehicle applications: Issues and challenges. *Renewable and Sustainable Energy Reviews*, *69*, 771–789.

15. Burke, A. (2007). R&D considerations for the performance and application of electrochemical capacitors. *Electrochimica Acta*, *53*(3), 1083–1091.

16. Wang, Q., Nie, Y. F., Chen, X. Y., Xiao, Z. H., & Zhang, Z. J. (2016). Controllable synthesis of 2D amorphous carbon and partially graphitic carbon materials: Large improvement of electrochemical performance by the redox additive of sulfanilic acid azochromotrop in KOH electrolyte. *Electrochimica Acta*, *200*, 247–258.

17. Duffy, N. W., Baldsing, W., & Pandolfo, A. G. (2008). The nickel–carbon asymmetric supercapacitor—Performance, energy density and electrode mass ratios. *Electrochimica Acta*, *54*(2), 535–539.

18. Faggioli, E., Rena, P., Danel, V., Andrieu, X., Mallant, R., & Kahlen, H. (1999). Supercapacitors for the energy management of electric vehicles. *Journal of Power Sources*, *84*(2), 261–269.

19. Muzaffar, A., Ahamed, M. B., Deshmukh, K., & Thirumalai, J. (2019). A review on recent advances in hybrid supercapacitors: Design, fabrication and applications. *Renewable and Sustainable Energy Reviews*, *101*, 123–145.

20. Chen, X., Paul, R., & Dai, L. (2017). Carbon-based supercapacitors for efficient energy storage. *National Science Review*, *4*(3), 453–489.

21. Rakhi, R. B., Alhebshi, N. A., Anjum, D. H., & Alshareef, H. N. (2014). Nanostructured cobalt sulfide-on-fiber with tunable morphology as electrodes for asymmetric hybrid supercapacitors. *Journal of Materials Chemistry A*, *2*(38), 16190–16198.

22. Purushothaman, K. K., Saravanakumar, B., Babu, I. M., Sethuraman, B., & Muralidharan, G. (2014). Nanostructured CuO/reduced graphene oxide composite for hybrid supercapacitors. *RSC Advances*, *4*(45), 23485.

23. Zhao, N., Deng, L., Luo, D., & Zhang, P. (2020). One-step fabrication of biomass-derived hierarchically porous carbon/MnO nanosheets composites for symmetric hybrid supercapacitor. *Applied Surface Science*, *526*, 146696.

24. Chebrolu, V. T., Balakrishnan, B., Selvaraj, A. R., & Kim, H.-J. (2020). A core–shell structure of cobalt sulfide//G-ink towards high energy density in asymmetric hybrid supercapacitors. *Sustainable Energy & Fuels*, *4*(9), 4848–4858.

25. Wan, L., Liu, J., Li, X., Zhang, Y., Chen, J., Du, C., & Xie, M. (2020). Fabrication of core-shell NiMoO4@MoS2 nanorods for high-performance asymmetric hybrid supercapacitors. *International Journal of Hydrogen Energy, 45*(7), 4521–4533.

26. Zhao, Y., Liu, H., Hu, P., Song, J., & Xiao, L. (2020). Asymmetric hybrid capacitor based on Co_3O_4 nanowire electrode. *Ionics, 26*(12), 6289–6295.

27. Machida, K., Suematsu, S., Ishimoto, S., & Tamamitsu, K. (2008). High-voltage asymmetric electrochemical capacitor based on polyfluorene nanocomposite and activated carbon. *Journal of the Electrochemical Society, 155*(12), A970.

28. Opportunity Charged Electric Buses _ Environmental blog.html. (n.d.). https://volvo-busesenvironmentblog.wordpress.com/2015/11/17/opportunity-charged-electric-buses/

29. Shin, J., Shin, S., Kim, Y., Ahn, S., Lee, S., Jung, G., … Cho, D.-H. (2014). Design and implementation of shaped magnetic-resonance-based wireless power transfer system for roadway-powered moving electric vehicles. *IEEE Transactions on Industrial Electronics, 61*(3), 1179–1192.

30. Dong, L., Ma, X., Li, Y., Zhao, L., Liu, W., Cheng, J., … Kang, F. (2018). Extremely safe, high-rate and ultralong-life zinc-ion hybrid supercapacitors. *Energy Storage Materials, 13*, 96–102.

31. Zhang, L., Hu, X., Wang, Z., Sun, F., & Dorrell, D. G. (2018). A review of supercapacitor modeling, estimation, and applications: A control/management perspective. *Renewable and Sustainable Energy Reviews, 81*, 1868–1878.

32. Al, M., Gualous, H., Omar, N., & Van, J. (2012). Batteries and supercapacitors for electric vehicles. In Z. Stevic (Ed.), *New Generation of Electric Vehicles*. InTech.

33. Grunditz, E., & Jansson, E. (n.d.). *Modelling and Simulation of a Hybrid Electric Vehicle for Shell Eco-marathon and an Electric Go-kart*, Chalmers University Library, *182*.

34. Sedlakova, V., Sikula, J., Majzner, J., Sedlak, P., Kuparowitz, T., Buergler, B., & Vasina, P. (2015). Supercapacitor equivalent electrical circuit model based on charges redistribution by diffusion. *Journal of Power Sources, 286*, 58–65.

35. Logerais, P. O., Camara, M. A., Riou, O., Djellad, A., Omeiri, A., Delaleux, F., & Durastanti, J. F. (2015). Modeling of a supercapacitor with a multibranch circuit. *International Journal of Hydrogen Energy, 40*(39), 13725–13736.

36. Udhaya Sankar, G., Ganesa Moorthy, C., & RajKumar, G. (2019). Smart storage systems for electric vehicles: A review. *Smart Science, 7*(1), 1–15.

11 Power Harvesting and Storage System

Supercapacitors Aiding New and Renewable Energy Generation

Nijisha P. and Shidhin M.

CONTENTS

11.1 INTRODUCTION

Energy harvesting, also known as power harvesting, uses renewable energy to power small electronic or electrical devices. An energy harvester that can produce energy in every circumstance and its integration with a storage device is fundamental to meet the increasing energy demands of modern society.

Among the renewable energy resources photovoltaic (PV) power generation is the most attractive energy harvesting solution. The average power density of solar radiation impinging on the earth's surface is sufficient to charge the devices in a short time from everywhere. But PV power generation depends highly on weather conditions, making solar power intermittent and is available only for few hours. This fluctuation in generation can be overcome by using a suitable energy storage device. Rapid changes in PV power due to rapid changes in illumination conditions can be smoothed using energy storage devices by delivering power when the solar supply is scarce [1, 2].

Wind turbines are another widely utilized renewable energy resource. The amount of energy produced by a wind turbine is proportional to its rotational speed. However, due to the continuous change in wind speed, consistent power production from wind turbines is never attained, and they cannot be directly linked to the grid [3]. This problem can be solved by connecting an energy storage device to the turbine output in parallel. Despite the fact that blue energy is not widely employed in current power systems, substantial progress has been made in this field recently. To be more specific, the energy created as fresh water mixes with saline water is used in capacitive energy extraction by utilizing the salinity gradient [4]. In view of current climate change, these strategies are becoming increasingly important.

Supercapacitors [5–7], lithium batteries [8–11], lead-acid batteries [12], redox-flow batteries [13, 14], etc., are the commonly studied energy storage devices combined with PVs, wind turbines, blue energy, etc. Among these energy storage devices, supercapacitors have advantages over the abovementioned devices, like high energy density, ultrafast charging and discharging, wide operating temperature range, and long cycle life. So, the integration of PV cells with sustainable supercapacitors became the promising hybrid system for the harvesting and storage of renewable energy [5, 6]. This helps to reduce transmission losses and the variability in power output caused by solar radiation fluctuation [1].

11.2   INTEGRATED SOLAR CELL–SUPERCAPACITOR SYSTEM

The configuration of solar cells with supercapacitors can be done in two ways: by integrating the solar energy conversion part and the electrical energy storage unit either in one device or by sharing a common electrode as a connection. So, there must be a front electrode for the solar energy harvesting followed by conversion and a counter electrode for the supercapacitor. Upon light illumination, the photocharging process takes place. i.e., the photo anode harvests and converts the solar energy into electrical energy and charges the supercapacitor. The discharging process happens when the capacitor supply power is connected to an external load [1, 2, 15, 16].

11.2.1   Device Architecture

The first prototype of a hybrid device for solar energy conversion and storage was introduced in 1970 [17]. It was a two-electrode planar integrated device, and the electrodes were directly in contact with each other. The storage efficiency of such devices was low because of the large resistance loss as a result of contact between the storage electrode and other electrodes. Upon introducing a cation-specific membrane in between the electrodes, the integrated system yielded a storage efficiency of ~90%. According to the number of electrodes used in a planar integrated system, the devices are classified as two-electrode [18, 19], three-electrode [20, 21], and four-electrode [22, 23] units. Among them, the three-electrode unit monolithic structure is the most often reported architecture. Upon rolling up the planar structure gives rise to integrated fiber-shaped devices with common electrodes. The hybrid system looks like a core shell with a fiber shaped solar cell and supercapacitor part [24]. These fiber structures showed better photon absorption from reflected and scattered light and hence a better performance. A schematic of the conventional integration of the solar cell and supercapacitor, planar three-electrode configuration with common electrode, and a coaxial fiber structure is shown in Figure 11.1.

Silicon solar cells are the most commercially successful PV technology to date. These first-generation cells have attained an overall efficiency of 25% and have completely dominated the terrestrial PV market. However, they are rigid devices and hinder their application in flexible and portable electronics. After several years of research, third-generation solar cells including dye-sensitized solar cells Photo conversion Efficiency (Photo conversion efficiency (PCE) = 14.1%), organic solar cells (18.4%), perovskite solar cells (25.8%), quantum dots solar cells (18.1%), etc., were introduced. These new-generation solar cells can be fabricated on flexible substrates via low temperature processing and are compatible with weaving and ideal for portable and wearable electronics. This part of the chapter discusses the hybrid energy storage and conversion devices by integrating solar cells with supercapacitors. Among the three generations of solar cell, organic solar cells, silicon solar cells, dye-sensitized solar cells, etc., have been successfully coupled with supercapacitors for fabricating a self-powering system and these hybrid devices can be called "solar capacitors".

11.2.2   INTEGRATED SILICON SOLAR CELL–SUPERCAPACITOR SYSTEM

A silicon solar cell consists of p-type and n-type silicon layers placed one above the other. At the junction the electrons from the n-type side move into the p-type side where holes are present and form a depletion zone. The n-side of the depletion zone contains holes, and the p-side contains electrons. The presence of opposite charges creates an internal electric field. When solar radiation falls on silicon PV cells, the field will move electrons to the n-type side and holes to the p-type side and is collected at the electrodes to the external circuit [25, 26].

Pint et al. in 2014 developed a silicon supercapacitor [27]. This was done by integrating an active material for energy storage on the back of the commercially available PV device. The aluminum current collector of commercially available silicon PV device was dissolved using KOH followed by electrochemical etching using HF on the p-type side to the solar cell. To complete the hybrid structure, a PEO and 1-ethyl-3-methyl imidazolium tetrafluoroborate-based solid state electrolyte was sandwiched with a single crystal silicon counter electrode. The device was tested using a galvanostatic charge/discharge method with a current of 0.4 A/cm^2 and a cutoff voltage of 0.55 V, yielding an 84% coulombic efficiency and a total capacitance of 0.14 F/m^2. But etching on the rear side of the silicon solar cell may affect the current collection.

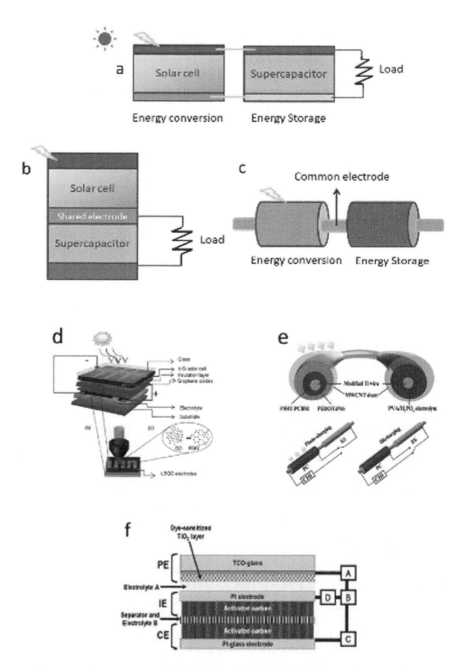

FIGURE 11.1 (a) Conventional integration of photovoltaic cell and supercapacitor, (b) planar or monolithic three-electrode structure, (c) coaxial fiber parallel structure, (d) schematic of an on-chip integrated energy storage system with crystalline silicon solar cell [28], (e) schematic of a coaxial flexible energy fiber [34], and (f) schematic of a three-electrode sandwich structure [39].

Min Gu and team reported a concept of an on-chip integrated energy storage system with a crystalline silicon solar cell employing a graphene oxide (GO) film laser-scribed on the rear side of the cell [28]. The structure of the device is shown in Figure 11.1(d). A CO_2 laser beam was used to scribe GO on the commercial silicon solar cell. The laser beam induced the conversion of GO to rGO. Defects were generated in the graphene layer during laser scribing resulting in the formation of porous structures. The conductivity measurement of laser-scribed GO film with laser power showed an increase in trend with a maximum value of 10^4 S/m. To avoid the problem of leakage and volatilization, an ionogel electrolyte was utilized. A layer of SU-8 was used as insulation between the solar cell and the supercapacitor part in order to avoid the impact from scribing.

From the individual analysis done on the solar cell part and supercapacitor part, it was found that an improvement in the solar cell performance was found due to the anti-reflection layer on the rear side. For the supercapacitor, a coulombic efficiency of 68% was observed with a V_{oc} of 0.38 V. The device has shown a long self-discharge time. But the output of the supercapacitor was very low when compared to the V_{oc} of the solar cell and the main loss was due to the contact resistance at the current collection tape and the Laser-scribed Graphene Oxide (LSGO). The device showed an energy density of 51 Wh cm^{-3} and the power density is around 4.6 Wcm^{-2}. The self-discharge study performed showed that the charge was detained for more than 10 days which is not common in commercial supercapacitors.

11.2.3 Integrated OSC–Supercapacitor System

An organic solar cell (OSC) commonly known as a polymer or plastic solar cell consists of thin layers of organic semi-conducting materials [29, 30]. A photoactive layer consisting of two different types of semi-conductors – donor and acceptor material is sandwiched between two thin film electrodes. This photoactive material forms the hetero-junction in inorganic solar cells. When light is impinged on the semi-transparent electrode, electrons get ejected from the photoactive donor material leaving behind a hole. These two charge carriers immediately bind together to form an electron–hole pair known as exciton. The exciton then moves towards the interface of donor-acceptor and gets separated into charge carriers. These charges get collected at the respective electrodes and then to the external circuit. OSCs with the highest efficiency of 18.4% were reported in 2021 [31]. The fabrication of OSCs on flexible substrates was possible via low temperature processing. Also, the lightness and printability makes them promising candidates among different generations of solar cells.

By incorporating OSCs and supercapacitors, a printable photo-supercapacitor was introduced by M. Srinivasan using single-walled carbon nanotube (CNT) [32]. The photoanode for the solar cell part is made of spin coating poly(3,4-ethylenedioxythiophene) poly(styrenesulfonate) (PEDOT:PSS) on ITO substrate. On top of this a mixture of poly(3-hexylthiophene) and [6,6]-phenyl-C61-butyric acid methyl ester is spin-coated to form the photoactive material. Aluminum cathodes are deposited by thermal evaporation. To integrate the supercapacitor with the OSC, a CNT solution drop-casted on the aluminum cathode functions as one of the electrodes for symmetric supercapacitors. A free standing PVA-phosphoric acid electrolyte was then sandwiched between two CNT electrodes to obtain the device. Upon illuminating the photoanode side of the device, P3HT absorbs the light energy and excitons are produced. The work function difference between the electrodes acts as the driving force that helps the holes to migrate towards the anode and electrons towards the cathode. The hybrid system, with a common integration platform, results in a thinner and lighter device. When the two devices – supercapacitor and OSC – were externally connected using wire, the internal resistance was found to be 220 Ω and for the design discussed above, the internal resistance was found to be 115 Ω. i.e., a 43% decrease in internal resistance was obtained. However, the energy storage capacity of a photosupercapacitor was lower. Two OSCs were linked in series to enhance the charging voltage to get maximum storage capacity to the supercapacitor. A specific capacitance value of 79.8 F g^{-1} was observed for this modified hybrid system.

An indoor optimal photo-rechargeable system was developed by Arias et al. with a total efficiency of 2.9% which is sufficient to power low consumption electronic devices [33]. PEDOT:PSS acts as the hole transport layer, polyethylenimine ethoxylated (PEIE) as the electron transport layer with PCDTBT:PC71BM as the active layer. Under 1 sun illumination, a PCE value of 6.2% was attained and cells optimized for indoor conditions attained 7.6% of PCE. Upon integrating the OSC with a fully printed supercapacitor, a total energy conversion and storage efficiency of 1.7% was achieved under 1 sun (26 mJ energy and 4.1 mW power). Under simulated indoor conditions, an efficiency of 2.9% with an energy value of 13.3 mJ and a power of 2.8 mW were obtained. To enhance the efficiency of the OSC under indoor light, the dark current of the cell has to be minimized.

Peng et al. introduced an all-solid-state flexible energy fiber for energy conversion and storage purposes [34]. The coaxial structure favors fast charge transport and the use of flexible and transparent multi-walled carbon nanotubes (MWCNTs) as electrodes provide effective photoelectric conversion and energy storage. In this configuration, titania nanotube-modified Ti wire acts as one of the electrodes and aligned MWCNT acts as the second electrode for the supercapacitor and OSC. Poly(3-hexyl thiophene):-phenyl-C 61-butyric acid methyl ester and poly(3,4-ethylenedioxythiophene) forms the photoactive layer in the OSC. Poly(styrene sulfonate) (PEDOT:PSS) and PVA-H$_3$PO$_4$ gel acts as the electrolyte in OSC and supercapacitor respectively. A schematic of the hybrid device is shown in Figure 11.1(e).

The efficiency of the energy fiber can be calculated by multiplying the photoenergy conversion efficiency in the OSC and the energy storage efficiency in the supercapacitor part.

$$\eta = \eta_{conversion} * \eta_{storage}$$
$$= E_{ES} * S_{ES} = E_{SC} * S_{SC}/\left(P_{in} * t * S_{PC}\right)$$

(11.1)

where E_{SC}: energy storage efficiency, S_{SC}: surface area of the ES part, P_{in}: illuminated light-energy density, t: photocharging time, and S_{PC}: effective area in the OSC part respectively.

The device showed a high electrochemical stability and from the galvanostatic charging/discharging curve, the specific capacitance with respect to length was calculated to be 0.77 mF cm^{-1} with 1.61 * 10^{-7} Wh cm^{-2} of energy density. A study on the dependence of the MWCNT thickness on efficiency study showed that with increase in thickness of the MWCNT layer from 0.5, 2, 5, and 10 to 20 μm, the efficiency of the energy fiber increased from 0.20, 0.35, 0.40, and 0.66 to 0.82%. Upon further increasing the thickness, the efficiency remained unchanged. Upon bending this energy fibers flexible energy fiber for 1000 cycles, they showed only a slight decrease in efficiency of less than 10%. These flexible structures can be woven with fibers to form flexible textiles.

11.2.4 INTEGRATED DSSC–SUPERCAPACITOR SYSTEM

Michael Gratzel and Brian O'Regan demonstrated the first dye-sensitized solar cells (DSSCs) using titania nanocrystals [35]. In DSSCs, the photoreception and charge transport were implemented by different components which are contrary to a conventional PV cell. A DSSC consists of a sandwich structure with a photoanode and counter electrode with electrolyte in between. The photoanode is typically a transparent conducting oxide (TCO) coated with a mesoporous oxide layer sensitized with a monomolecular layer of organic or organometallic dye. The electrolyte is a redox couple that helps in the regeneration of dye and the electrolyte during the operation and the counter electrode is a TCO coated with a catalyst [36, 37].

A DSSC makes a suitable energy conversion unit because of its easy fabrication, low cost, its ability to harvest energy under diffuse light, and flexibility [36, 37]. But the converted energy has to be successfully stored and the supercapacitors are attracting considerable attention in this area because of their high power density and durability when compared to batteries. The simple charge storage mechanism, generation of less heat, and use of aqueous electrolyte promise sustainable charge/discharge cycles before decay. Thus, the integration of DSSC and supercapacitor devices for the harvesting, conversion, and storage of energy are a great choice.

The first self-charging capacitor with the help of solar radiation was introduced by Miyasaka and is named a *photocapacitor* [38]. It was a two-electrode sandwich-like structure consisting of a photoanode, a liquid electrolyte (redox free), and a counter electrode with an area confined to 0.64 cm^2. The photoanode is a Ru dye-sensitized TiO$_2$ layer on a TCO substrate. A hole-trapping layer was added at the semiconductor nanoparticle–activated carbon interface. Platinum-deposited TCO, onto which a layer of activated carbon is coated, acts as counter electrode. The gap between the electrodes is filled with electrolyte solution. The photogenerated positive and negative charges are stored on the surface of activated carbon on the photoanode and cathode as a double layer respectively. The hybrid device with a charging voltage of 0.45 V achieved a capacitance of 0.69 F cm^{-2} per area. Upon discharging, the cell voltage exhibited a constant decrease with time, by an initial internal resistance drop. In this configuration, a high internal resistance is reported which retards the discharge process. To avoid this loss a three-electrode configuration was demonstrated by introducing an internal bifunctional electrode between the working and counter electrodes [39]. The cell structure consists of a photoanode (Ru dye-coated TiO_2/ TCO substrate), an inner electrode which consists of Pt-plate coated with activated carbon, and a counter electrode (Pt-coated glass covered with activated carbon) as shown in Figure 11.1(f). The internal electrode is a junction that acts simultaneously as cathode and anode for the outer electrodes. This internal electrode stores charges on one side and conducts the redox electrons on other side.

A capacitance of 0.65 F cm^{-2} was obtained with a small decrease in the voltage during the beginning of discharge. The internal resistance was reduced to 330 Ω for a three-electrode hybrid cell when compared to the two-electrode system having 2.6 kΩ. An enhancement in performance was due to the efficient transfer of electrons and holes at the bifunctional region in the inner electrode. The energy densities of discharge per unit area of 47 mW h cm^{-2} and a charging voltage up to 0.8 V was obtained with this configuration which was better than the two-electrode photoca-pacitor. However, a low coulombic efficiency was reported due to the back-electron transfer at the inner electrode resulting in the quenching of holes collected at the activated carbon by iodide anions.

The conducting polymers are widely used as supercapacitor materials because of their excellent specific capacitance. Among them, PEDOT showed the highest elec-trode-specific capacitance approaching 5 F cm^{-2}. But its mass specific capacitance is relatively small. A flexible plastic photo-rechargeable capacitor with a common Pt electrode and (poly(3,4-ethylenedioxythiophene) (PEDOT) polymer films as the supercapacitor material was introduced by Kuo-Chuan Ho [40] with a specific capac-itance of 0.52 F cm^{-2}. Also, the high conductivity and surface area of the PEDOT-conducting polymer leads to a lower internal resistance of 160 Ω. This is about half the internal resistance reported for the first three-electrode devices reported by Miyasaka. The light harvested electrons flow from the DSSC to the supercapacitor part where it gets stored in the PEDOT layer. The same group used poly(3,3-diethy l-3,4-dihydro-2thieno-[3,4-b][1,4]dioxepine) – PProDOT-Et2, another derivative of poly(3,4- alkylenedioxythiophene)s – (PXDOTs), and the photocapacitor showed a photocharged voltage of 0.75 V, an area-specific capacitance of 0.48 F/cm^2 with an energy storage efficiency of 0.6%. The diethyl substituent of the polymer cre-ated enough space for the facile movement of the ions during the redox process that enhanced the redox property and stability of the device.

An integrated flexible power fiber that incorporates a DSSC and supercapacitor was introduced by Zou et al. They used a polyaniline-coated stainless steel wire as the electrode which has a dual function: as the counter electrode for fiber DSSCs and electrode for fiber supercapacitors. Fiber supercapacitors consist of two parallel working electrodes made of polyaniline-coated stainless steel wire and in order to prevent short circuits from occurring between the two electrodes, one of the working electrodes is surrounded by a helical space wire at a certain pitch. The photoanode of fiber-DSSC consist of Ti wire coated with TiO_2 sensitized with dye. A switch exists between the photoanode and the working electrode of the supercapacitor. Upon illu-mination the electrons are stored in the supercapacitor and during discharge the switch is manually disconnected. The integrated system attained a photocharged voltage of 0.621 V and the total energy conversion of the power fiber is 2.1%.

Peng et al. introduced an integrated device using MWCNTs and MWCNT-polyaniline composite films as electrodes in integrated device [41]. The device used well aligned and free-standing MWCNT films as electrodes and exhibited an overall conversion and storage efficiency of 5.12%. The high performance of the device was attributed to the aligned structure and high surface area of MWCNTs. The photo-voltage of the integrated device attained a value of 0.72 V. The specific capacitance of the device was found to be 48 F g^{-1}, and a storage efficiency of 84%. By incorporating polyaniline into MWCNTs, the specific capacitance of the device was further improved to 208 F g^{-1}.

The same group introduced a coaxial energy fiber that consisted of an aligned CNT sheet wrapped around an elastic rubber fiber which is then coated with a thin layer of gel electrolyte made of polyvinyl alcohol-H_3PO_4. It is then covered with another sheet of CNT electrode to form the energy storage component. The super-capacitor part is then inserted into an elastic tube. A third sheet of CNTs is wrapped around this plastic elastic tube that forms the cathode of the photo-conversion system. The entire assembly is then incorporated into a TiO_2 nanotube-grown helical Ti fiber which acts as the photoanode. The resulting device is inserted into a polyethyl-ene tube filled with the redox electrolyte. In this coaxial structure, the solar energy was converted in the sheath and the electrical energy was stored in the core. The photoelectric-conversion efficiency of the energy fiber is calculated to be 6.47% and the voltage of the storage component attained a value of 0.65 V. The high electro-catalytic activity of the aligned CNT sheets helped in the improved performance of the device. This flexible structure showed an overall conversion efficiency of 1.83% which is maintained even after stretching and bending.

11.3   WIND TURBINES

Wind power production systems distinguish themselves from other types of power facilities for their input power control strategy that employs a turbine drive actuator, whereas the active and reactive powers of typical fossil-fuel-based power plants are controlled by the fuel injection system at the inlet and the automated voltage regula-tion system, respectively. However, because of the varying wind velocity reliant on the climate, the traditional active power regulation approach is unable to employ in wind turbines. Furthermore, owing to rapid fluctuations in wind speed, the power output from the wind turbines is never certain, it may result in a slew of difficulties and system instability.

For power system applications, energy must be dissipated promptly without sacri-ficing power quality, especially in emergency power applications. On the other hand, for applications such as power management, including load curve levelling and peak shaving, the energy must be discharged for several hours. Large-scale energy storage systems may be implemented using a variety of approaches. Because all energy stor-age techniques are costly, economic calculations are very essential.

Non-flat power demand is one of the most critical challenges in power sys-tems. This mandates the use of backup power production technologies to supple-ment the power grid during peak hours. The issue may be addressed by utilizing storage devices, which can also contribute to an increase in clean energy usage, improved active and reactive power regulation, a reduction in voltage fluctuations, and improved power quality and system stability [42]. Supercapacitor-based energy storage methods are described in the following sections, with a focus on storage for wind energy application.

11.3.1 WIND TURBINE POWER CHARACTERISTICS

In terms of wind speed, the power characteristic of the wind energy installation equates to the mechanical performance characteristic of the wind energy installation. To acquire this characteristic, first, we need to compute the mechanical power of the wind turbine.

The kinetic energy (E_k) in the wind is calculated as follows:

$$E_k = \frac{1}{2}mv_\omega^3 \tag{11.2}$$

According to Equation (11.1), wind power is equal to:

$$P_\omega = \frac{dE_k}{dt} = \frac{1}{2}\rho A V_\omega^3 \tag{11.3}$$

where P_ω is wind power, ρ is air density in kg/m³, A is sweep area in m², m is object mass in kg, and V_ω is the linear wind speed in m/s [43]. Even though Equation (11.2) describes wind power, the power transmitted to the wind turbine is decreased by the power coefficient (C_p). As a result, the following is the connection between wind power and mechanical power supplied to the turbine shaft is:

$$P_m = C_p P_\omega \tag{11.4}$$

C_p is affected by wind speed, turbine rotor rotation speed, and the angular position of the turbine rotor blades (β). According to the Betz limitations, this coefficient has a theoretical maximum value of 0.59 [42]. In other words, in principle, around 59% of the kinetic energy of the wind may be turned into mechanical energy. The ideal efficiency factor for modern three-bladed wind turbines is in the region of 0.45–0.55. C_p is represented as a function of tip speed ratio (λ) and pitch angle in different designs (β). The tip speed ratio is denoted by the following equation:

$$\lambda = \frac{r_T \omega_m}{\upsilon_\omega} \tag{11.5}$$

where ω_m and r_T are the rotational speed of the blade and the radius of the turbine rotor, respectively. The mechanical power supplied to the turbine shaft is expressed as follows by combining Equations (11.4) and (11.5):

$$P_m = \frac{1}{2}\rho A C_p(\lambda, \beta) V_\omega^3 \tag{11.6}$$

For the majority of wind turbines, the power coefficient is represented as Equations (11.7) and (11.8):

$$C_p(\lambda, \beta) = c_1 \left\{ \frac{c_2}{\lambda_i} - c_3\beta - c_4 \right\} e^{\frac{c_5}{\lambda_i}} + c_6\lambda \tag{11.7}$$

$$\frac{1}{\lambda_i} = \frac{1}{\lambda + 0.08\beta} - \frac{0.035}{\beta^3 + 1} \tag{11.8}$$

11.3.2 SUPERCAPACITORS LINKED TO WIND FARMS

Supercapacitors are used in wind turbines to keep the DC connection voltage stable and to reduce output power variations. To put it another way, a supercapacitor in the electrical output from a wind turbine compensates for power variations caused by irregularities in the wind speed. The electrical structure of supercapacitors used in wind turbines is made up of a supercapacitor bank and a two-switch DC/DC converter that is linked to a Doubly-fed Induction Generator (DFIG) through a DC connection [42]. The active power output is regulated using a buck-boost converter. For this, reference power is initially identified, when the network power exceeds the reference power, the supercapacitor charges, and the converter enters buck mode. If the network power is less than the reference power, the supercapacitor discharges and the DC/DC converter operates in boost mode. The capacity of the supercapacitor bank is computed as:

$$C_{ess} = \frac{2P_n T}{v_{sc}^2} \tag{11.9}$$

where C_{ess} is in farads, P_n is the rated power of the DFIG in watts, V_{sc} is the voltage rating of the supercapacitor bank in volts, and T is the intended period in seconds that the energy storage system can supply/store energy at the rated power of the DFIG.

11.4 BLUE ENERGY: CAPACITIVE STORAGE

11.4.1 INTRODUCTION

There are countries (like India) with a large number of rivers that flow throughout the country and blend into the ocean. This geographical advantage can be effectively utilized to generate clean and pollution-free electricity. The salinity difference between sea and river water has recently been identified as a possible source of clean energy known as blue energy or osmotic energy [44]. When a river flows into the sea, the spontaneous and irreversible mixing of fresh and salt water raises the entropy of a system. A portion of the entropy shift can be translated from fluid thermal energy to electrical energy [45]. It is estimated that about 0.8 kWh m^{-3} can be produced as a function of this entropy change at the sea/river interface. Various methods for extracting energy from a salinity gradient have been suggested, including pressure-retarded osmosis (PRO), reverse electro-dialysis (RED), concentration electrochemical cells, and devices that exploit differences in vapor pressure [46–50]. Researchers have recently investigated electrochemical capacitance-based blue energy extraction technologies.

11.4.2 THEORETICAL ANALYSIS

The energy that can be extracted from blue energy by combining river and sea water can be better understood if the reverse process "distillation" is considered. In distillation, energy is provided to acquire freshwater from sea water. So theoretically the reverse process of distillation should potentially liberate energy [51, 52]. As a result, the Gibbs energy of the mixture $\Delta_{mix}G$ represents the non-expansion work generated by blending a concentrated brine solution s (sea water) with a dilute brine solution r (river water) with constant pressure p and absolute temperature T to form a brackish solution m [53–55]:

$$\Delta_{mix}G = G_m - \left(G_s + G_r\right) \tag{11.10}$$

If both solutions are completely diluted, the Gibbs energy of the blending process may be computed from the variance in molar entropy (i.e., $\Delta_{mix}H = 0$):

$$\Delta G = -\left(n_s + n_r\right)T\Delta_{mix}S_m - \left(-n_sT\Delta_{mix}S_s - n_rT\Delta_{mix}S_r\right) \tag{11.11}$$

where n denotes the quantity (moles) and $\Delta_{mix}S$ denotes the role of mixing molar entropy to the overall molar entropy of blending electrolyte solution:

$$\Delta_{mix}S = -R\sum_i C_i \ln C_i \tag{11.12}$$

R stands for the universal gas constant, which is 8.314 J/mol K, and c stands for the mole fraction of component i (I = Na$^+$, Cl$^-$). The change in Gibbs free energy among mixed and individual solutions prior mixing may be computed as a function of the volumetric fraction of sea water, x, as illustrated below:

$$\Delta_{mix}G = 2RT\left[C_m \ln(C_m) - xC_s \ln(C_s) - \left(1 - x\right)C_r \ln(C_r)\right] \tag{11.13}$$

The maximal amount of energy that may be extracted from surplus saltwater is $\Delta G \approx 2500$ JL^{-1} of freshwater ($x \approx 1$). This indicates that a power station treating 400 m^{3}S^{-1} of freshwater every second might generate up to 1 GW of electricity. The world's biggest river (the Amazon) has an annual average flow rate of 179,000 m^3/s, theoretically, this process will create as much as 437.5 GW when it reaches the sea which is 19 times that of The Three Gorges Dam, the world's biggest hydroelectric facility.

11.4.3 CAPACITIVE ENERGY EXTRACTION: ELECTRIC DOUBLE LAYER

"Capacitive energy extraction" is a novel and intriguing approach for converting the energy generated by combining fresh and salt water into electricity. This relatively new capacitive energy extraction method is rooted in the electrochemical double layer (EDL) capacitor technology. The Gouy–Chapman–Stern model presently describes EDL, which represents the ions spreading near to the electrodes as the combination of an adsorbed EDL and a diffuse EDL [56–59]. Charged ions in the diffuse EDL attain an equilibrium amongst diffusion, which attempts to equalize ion concentration and electrostatic force, which tends to exacerbate charge imbalance near to the surface. The charge is entirely shielded at a considerable distance from the electrode; thus, the electric field is only present within the diffuse EDL [55, 58].

The following equation defines the relationship between the potential difference φ between the electrode, the surface charge density s, and the bulk solution:

$$\varphi = \frac{2k_B T}{e} \sinh^{-1}\left(\frac{\sigma}{\sqrt{8CN_A \varepsilon_0 \varepsilon_r k_B T}} \right) \qquad (11.14)$$

The equation holds for asymmetric, monovalent electrolytes like NaCl, where k_B is the Boltzmann constant, T is the temperature, e is the electron charge, N_A is the Avogadro constant, ε_0 is the electric constant, ε_r is the relative dielectric constant, and C is the electrolyte concentration. According to Equation (3.5), the connection between charge and voltage is depicted in Figure 11(a). A reduction in concentration results in a rise in potential. While an EDL capacitor composed of active carbon electrodes is submerged in brine, the charges are stored in the EDLs, which are composed of counter ions and electrons distributed near the electrode/electrolyte interface.

Brogioli developed an EDL capacitor in 2009 to harvest energy from the salinity of the water. He suggested using porous carbon electrodes to preferentially combine liquids in a four-stroke charge–desalination–discharge–resalination cycle, akin to a Stirling thermal engine operating an expansion–cooling–compression–heating process [45]. The cell has two electrodes and is loaded with the changeable editable liquid, as shown in Figure 11.2(b). Both water tanks with dissimilar concentrations of NaCl, which stand for saltwater and freshwater, are linked to the cell via pumps that are triggered successively, such that the concentration of the fluid inside the cell can vary from low to high and vice versa. In Brogioli's work, highly porous and conductive electrodes of activated carbon are shaped into discs with a diameter of 2 mm and a thickness of 0.1 mm. The switch can swiftly switch the electrodes between charging and discharging modes. The resistor of 1 kΩ represents the load, through which the charging and discharging currents travel.

The second version of the lab-scale Brogioli EDL capacitor, which includes dense graphite current collectors (250 μm) porous carbon electrodes (270 μm, porosity 65%), and open-mesh polymer spacer (250 μm). Before shredding the complete stack of all layers, all materials are chopped into 6 × 6 cm² pieces and assembled. The

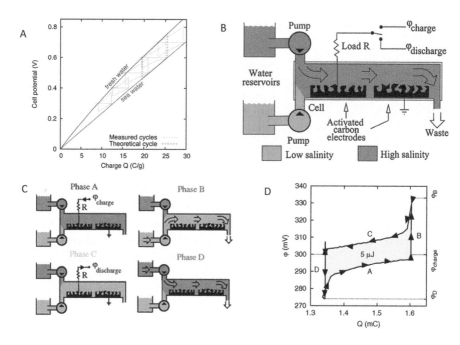

FIGURE 11.2 (a) Charge-voltage cycle at different EC-voltages V0, represented by dotted lines. The area enclosed by each cycle represents the extracted electrical energy from switching between 1 mM and 500 mM salt solutions. The lines labelled with "sea water" and "fresh water" represent the charge-voltage relations obtained from GCS theory, respectively, for 500 mM and 1 mM with $\eta = 0.99$ [56]. (b) Schematic view of the device. The cell contains two electrodes made of porous activated carbon, constituting a capacitor, which can be charged and discharged. The cell is filled with a NaCl solution, coming from one of the reservoirs. (c) Representation of the cycle on the potential versus charge graph. The area represents the extracted energy: in this case, it is 5 μJ. (d) Cycle for extracting energy from salinity difference. (a) Scheme of the four phases of the cycle [58].

aqueous NaCl solution is driven into a tiny hole (1.5×1.5 cm^2) in the center of the heap, where it runs via the spacer channel and exits the cell on all four sides. In all tests, the overall flow rate of 1 mL s^{-1} remained the same. The circuit contains a potentiostat that simulates the EDLC and operates with a voltage V_0 and a resistor of 11 Ω which acts as a load and is connected with the flow cell [57].

The extracted energy in the experiment with the second-generation configuration is roughly 2 J for EC voltages of $V_0 > 0.5$ V, which is 10^6 times the quantity extracted from the microscale prototype previously [18, 20]. Though the energy extraction performance is far from real, it offers a solid foundation for future improvement. Furthermore, if the solution cycle speed of dissimilar frequencies can be obtained, the power generation will get diverse orders to meet various applications [60]. By appropriately structuring the structure of the porous electrode material, the energy and power performance of EDL capacitors may be increased even more.

11.4.4   Capacitive Energy Extraction: Faradaic Pseudocapacitor

A pseudocapacitor is built on the charge accumulation given by the Faraday charge transfer mechanism of the capacitive electrodes [61, 62]. The pseudocapacitor is made up of two reversible electrochemical systems, the reactants of which are electrolyte salts and electrodes. Ions are stored on the electrodes during this process [63, 64]. The pseudocapacitor has a greater energy density and is less complicated to build than EDL capacitors [65, 66]. Cui introduced a unique pseudocapacitor based on Equation (11.15) to extract free energy from the salinity of the water:

$$5MnO_2 + 2Ag + 2NaCl \leftrightarrow Na_2Mn_5O_{10} + 2AgCl \tag{11.15}$$

Two different electrodes are used in this device: an anionic silver electrode, which interacts selectively with Cl⁻; it is a cationic electrode of MnO_2, which interacts selectively with Na⁺. These electrochemical reactions at two electrodes can be written as follows:

$$5MnO_{2(\alpha)} + 2Na^+_{(\varepsilon)} + 2e^- \leftrightarrow 2Na_2Mn_5O_{10} \tag{11.16}$$

$$2AgCl_{\beta'} + 2e^- \leftrightarrow 2Ag_{\alpha'} + 2Cl^-_\varepsilon \tag{11.17}$$

where α is the $Na_{2-x}Mn_5O_{10}$ phase, ε the electrolytic phase, β' the AgCl phase, and α' the silver phase. With respect to the typical hydrogen electrode (NHE), the potential of the two processes is given by:

$$E_+ = E_{+,0} + \frac{RT}{F}\ln\left[\frac{a_{Na,\varepsilon}}{a_{Na,\alpha}}\right] \tag{11.18}$$

$$E_- = E_{-,0} - \frac{RT}{F}\ln\left[a_{Cl,\varepsilon}\right] \tag{11.19}$$

where E_+ and E_- are the electrode potentials, $E_{+,0}$, and $E_{-,0}$ are the standard electrode potentials, $a_{Na,\alpha}$, $a_{Na,\varepsilon}$, $a_{Cl,a}$ are the activity of sodium in the solid phase (ε), and sodium ions in the electrolyte (α), and chloride ions in the electrolyte (α) respectively and ΔE is the potential difference between the two. If the sodium activity is constant in the solid phase (no current), then:

$$\Delta E = \Delta E_{0,a_{Na;a}} + 2\frac{RT}{F}\ln\left[C_{NaCl}\right] + 2\frac{RT}{F}\ln\left[\gamma_{NaCl}\right] \tag{11.20}$$

where ΔE_0 is the standard cell voltage, $CNaCl$ is the concentration of NaCl, and γ NaCl is the mean activity coefficient of NaCl, where A and B are the constants,

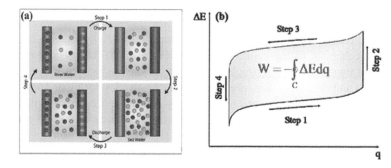

FIGURE 11.3 (a) Schematic representation of the working principle behind a complete cycle of the mixing entropy battery, showing how energy extraction can be accomplished: Step 1, charge in river water; Step 2, exchange to sea water; Step 3, discharge in sea water; Step 4, exchange to river water. (b) Typical form of a cycle of battery cell voltage (ΔE) vs charge (q) in a mixing entropy battery, demonstrating the extractable energy [54].

according to Debye-Hückellaw, γ NaCl dependency on CNaCl can be stated as follows:

$$\ln\left[\gamma_{NaCl}\right] = -\frac{A\sqrt{c_{NaCl}}}{1 + B\sqrt{c_{NaCl}}} \qquad (11.21)$$

As seen in Equation (11.21), increasing the NaCl concentration raises the ΔE. As a result, in Cui's device electrodes are charged in a low ionic strength solution (such as river water) after the initial removal of the Na^+ and Cl^- ions from the electrodes, corresponding Step 1 of Figure 11.3(a). The lesser ionic strength electrolyte is subsequently replaced by a concentrated electrolyte (sea water), causing a rise in cell open potential (Figure 11.3(a), Step 2). As the anions and cations are reincorporated into their respective electrodes, the "salinity pseudocapacitor" discharges at a greater open potential (Figure 3(a), Step 3). The high ionic strength solution is then withdrawn and replaced with the diluted electrolyte (river water), reducing the potential difference between the electrodes (Figure 11.3(a), Step 4). The interchange of solutions might take place through a powerful process, which could be useful for large-scale power generation. The energy of water salinity can be extracted in the cycle depicted in Figure 11.3(a), as in Figure 11.3(b) by a diagram of the expected shape of the voltage of the pseudocapacitor (E) as a function of the load (q) during a cycle in the closed cycle is during Steps 2 and 4 neither generates nor consumes energy. The capacitor requires energy in Step 1 to expel the ions from the crystal structure, whereas the capacitor creates energy in Step 3 by introducing the ions. As a result, the energy savings result from the fact that the same amount of charge is released at a greater voltage in Step 3 than was spent in Step 1.

$$W = -\oint_c \Delta E dq \qquad (11.22)$$

The geometric area of the electrode in contact with the solutions in the laboratory-scale salinity pseudocapacitor was 1 cm^2 with a 1-cm distance between the positive and negative electrodes. The pseudocapacitor was charged and discharged at 250 µA cm^{-2} for 20 minutes. The energy density generated by the prototype device is about 29 mJ cm^{-2} (power density 105 mW m^{-2}) in the case of sea water (0.6 M NaCl solution) and river water (0.024 M NaCl solution) [54].

Although the performance of a pseudocapacitor for energy extraction from water salinity is still far from practical application, with the optimization of the capacitor structure and materials engineering, the pseudocapacitor can be a promising technology that can make a significant contribution to renewable energy production.

11.5 SUMMARY

The chapter provides a brief review on self-powering units based on hybrid supercapacitor systems. A discussion on different designs, common materials used, working mechanisms, and performance of the integrated devices were presented. Solar cells of different generations, like silicon solar cells, OSCs, and dye-sensitized solar cells have been integrated with supercapacitors and the challenges encountered upon integration were listed. These hybrid varieties are highly desirable for next-generation electronics. Likewise, a discussion on the integration of wind turbines and blue energy was also presented. As discussed earlier, due to the variation in wind speed, the energy output from wind energy generation systems fluctuates. Therefore, to smoothen the output power, energy storage systems such as supercapacitors are a necessity before connecting to the grid. Blue energy generation using capacitive techniques is a relatively new method.

REFERENCES

1. V. Vega-Garita, L. Ramirez-Elizondo, N. Narayan, & P. Bauer, (2019) Integrating a photovoltaic storage system in one device: A critical review, *Prog Photovolt Res Appl.* 27 346–370.
2. A. Vlad, N. Singh, C. Galande, & P.M. Ajayan, (2015) Design considerations for unconventional electrochemical energy storage architectures, *Adv. Energy Mater.* 5 1402115.
3. P.H.A. Barra, W.C. de Carvalho, T.S. Menezes, R.A.S. Fernandes, & D. V Coury, (2021) A review on wind power smoothing using high-power energy storage systems, *Renew. Sustain. Energy Rev.* 137 110455.
4. C. Lian, C. Zhan, D. Jiang, H. Liu, & J. Wu, (2017) Capacitive energy extraction by few-layer graphene electrodes, *J. Phys. Chem. C.* 121 14010–14018.
5. N.A. Nordin, M.N.M. Ansari, S.M. Nomanbhay, N.A. Hamid, N.M.L. Tan, Z. Yahya, & I. Abdullah, (2021) Integrating photovoltaic (PV) solar cells and supercapacitors for sustainable energy devices: A review, *Energies.* 14 7211.
6. W.C. Chuang, C.Y. Lee, T.L. Wu, C.T. Ho, Y.Y. Kang, C.H. Lee, & T.H. Meen, (2020) Fabrication of integrated device comprising flexible dye-sensitized solar cell and graphene-doped supercapacitor, *Sens. Mater.* 32 2077–2087.
7. P.-O. Logerais, O. Riou, M.A. Camara, & J.-F. Durastanti, (2013) Study of photovoltaic energy storage by supercapacitors through both experimental and modelling approaches, *J. Solar Energy.* 2013 1–9.

8. H.-D. Um, K.-H. Choi, I. Hwang, S.-H. Kim, K. Seo, & S.-Y. Lee, (2016) Monolithically integrated, photo-rechargeable portable power sources based on miniaturized Si solar cells and printed solid-state lithium-ion batteries, *Energy Environ. Sci.* 9 1–10.

9. A. Vlad, N. Singh, J. Rolland, S. Melinte, P.M. Ajayan, & J.-F. Gohy, (2014) Hybrid supercapacitor-battery materials for fast electrochemical charge storage, Sci. Rep. 4 4315.

10. S.F. Hoefler, R. Zettl, D. Knez, G. Haberfehlner, F. Hofer, T. Rath, G. Trimmel, H. Martin R. Wilkening, & I. Hanzu, (2020) New solar cell–battery hybrid energy system: Integrating organic photovoltaics with Li-Ion and Na-Ion technologies, *ACS Sustainable Chem. Eng.* 8(51) 19155–19168.

11. Y. Di, S. Jia, X. Yan, J. Liang, & S. Hu, (2020) Available photo-charging integrated device constructed with dye-sensitized solar cells and lithium-ion battery, *New J. Chem.* 44 791–796.

12. G.J. May, A. Davidson, & B. Monahov, (2018) Lead batteries for utility energy storage: A review, *J. Energy Storage.* 15 145–157.

13. F. Urbain, S. Murcia-López, N. Nembhard, J. Vázquez-Galván, C. Flox, V. Smirnov, K. Welter, T. Andreu, F. Finger, & J.R. Morante, (2018) Solar vanadium redox-flow battery powered by thin-film silicon photovoltaics for efficient photoelectrochemical energy storage, *J. Phys. D Appl. Phys.* 52 (4) 044001.

14. P. Parmeshwarappa, R. Gundlapalli, & S. Jayanti, (2021) *Power and Energy Rating Considerations in Integration of Flow Battery with Solar PV and Residential Load Batteries,* 7 (3) 62.

15. D. Lau, N. Song, C. Hall, Y. Jiang, S. Lim, I. Perez-Wurfl, Z. Ouyang, & A. Lennon, (2019) Hybrid solar energy harvesting and storage devices: The promises and challenges, *Mater. Today Energy.* 13 22–44.

16. R. Liu, Y. Liu, H. Zou, & T. Song, Baoquan, (2017) Sun integrated solar capacitors for energy conversion and storage, *Nano Res.* 10(5) 1545–1559.

17. G. Hodes, J. Manassen, & D. Cahen, (1976) Photoelectrochemical energy conversion and storage using polycrystalline chalcogenide electrodes, *Nature.* 261 403–404.

18. A. Takshi, H. Yaghoubi, T. Tevi, & S. Bakhshi, (2015) Photoactive supercapacitors for solar energy harvesting and storage, *J. Power Sources.* 275 621–626.

19. T. Kanbara, K. Takada, Y. Yamamura, & S. Kondo, (1990) Photo-rechargeable solid state battery, *Solid State Ionics.* 4041 955–958.

20. Z. Ouyang, S.N. Lou, D. Lau, J. Chen, S. Lim, P.C. Hsiao, D.W. Wang, R. Amal, Y.H. Ng, & A. Lennon, (2017) Monolithic integration of anodic molybdenum oxide pseudocapacitive electrodes on screen-printed silicon solar cells for hybrid energy harvesting-storage systems, *Adv. Energy Mater.* 7 1602325.

21. Y.Y. Gui, F.X. Ai, J.F. Qian, Y.L. Cao, G.R. Li, X.P. Gao, & H.X. Yang, (2018) A solar rechargeable battery based on the sodium ion storage mechanism with $Fe_2(MoO_4)3$ microspheres as anode materials, *J. Mater. Chem. A.* 6 10627–10631.

22. N.F. Yan, G.R. Li, & X.P. Gao, (2014) Electroactive organic compounds as anode-active materials for solar rechargeable redox flow battery in dual-phase electrolytes, *J. Electrochem. Soc.* 161 736–741.

23. P. Liu, Y.L. Cao, G.R. Li, X.P. Gao, X.P. Ai, & H.X. Yang, (2013) A solar rechargeable flow battery based on photoregeneration of two soluble redox couples, *Chem- SusChem.* 6 802–806.

24. T. Chen, L.B. Qiu, Z.B. Yang, Z.B. Cai, J. Ren, H.P. Li, H.J. Lin, X.M. Sun, & H.S. Peng, (2012) An integrated "energy wire" for both photoelectric conversion and energy storage, *Angew. Chem. Int. Ed.* 51 11977–11980.

25. D.M. Chapin, C.S. Fuller & G.L. Pearson, (1954) A new silicon p-n junction photocell for converting solar radiation into electrical power, *J. Appl. Phys.* 25 676.

26. M.A. Green, (2009) The path to 25% silicon solar cell efficiency: History of silicon cell evolution, *Prog. Photovolt: Res. Appl.* 17 183–189.

27. A.S. Westover, K. Share, R. Carter, A.P. Cohn, L. Oakes, & C.L. Pint, (2014) Direct integration of a supercapacitor into the backside of a silicon photovoltaic device, *Appl. Phys. Lett.* 104 213905.

28. L.V. Thekkekara, B. Jia, Y. Zhang, L. Qiu, D. Li, & M. Gu, (2015) On-chip energy storage integrated with solar cells using a laser scribed graphene oxide film, *Appl. Phys. Lett.* 107 031105.

29. G. Chidichimo, & L. Filippelli, (2010) Organic solar cells: Problems and perspectives, 2010 1–11.

30. L.X. Chen, (2019) Organic solar cells: Recent progress and challenges, *ACS Energy Lett.* 4 2537–2539.

31. L.M. Svirskaite, S. Mandati, N. Spalatu, V. Malinauskiene, S. Karazhanov, V. Getautis, & T. Malinauskas, (2022) Asymmetric NDI electron transporting SAM materials for application in photovoltaic devices, *Synth. Met.* 291 117214.

32. G. Wee, T. Salim, Y.M. Lam, S.G. Mhaisalkar, & M. Srinivasan, (2011) Printable photo-supercapacitor using single-walled carbon nanotubes, *Energy Environ. Sci.* 4 413–416.

33. B.P. Lechêne, M. Cowell, A. Pierre, J.W. Evans, P.K. Wright, & A.C. Arias, (2016) Organic solar cells and fully printed super-capacitors optimized for indoor light energy harvesting, *Nano Energy.* 26 631–640.

34. Z. Zhang , X. Chen, P. Chen, G. Guan, L. Qiu, H. Lin, Z. Yang, W. Bai, Y. Luo, & H. Peng, (2013) Integrated polymer solar cell and electrochemical supercapacitor in a flexible and stable fiber format, *Adv. Mater.* 26(3) 466–470.

35. B. O'Regan & M. Grätzel, (1991) A low-cost, high-efficiency solar cell based on dye-sensitized colloidal TiO2 films, Nature 353 737–740.

36. J. Wu, Z. Lan, J. Lin,M. Huang, Y. Huang, L. Fan, & G. Luo, (2015) Electrolytes in dye-sensitized solar cells, *Chem. Rev.* 115(5) 2136–2173.

37. J. Gong, K.Sumathy, Q. Qiao, & Z. Zhou, (2017) Review on dye-sensitized solar cells (DSSCs): Advanced techniques and research trends, *Renew. Sustain. Energy Rev.* 68 234–246.

38. T. Miyasaka, & T.N. Murakami, (2004) The photocapacitor: An efficient self-charging capacitor for direct storage of solar energy, *Appl. Phys. Lett.* 85 3932–3934.

39. T.N. Murakami, N. Kawashima, & T. Miyasaka, (2005) A high-voltage dye-sensitized photocapacitor of a three-electrode system, *Chem. Commun.* 2005 3346–3348.

40. C.-Y. Hsua, H.-W. Chena, K.-M. Leeb, C.-W. Huc, & K.-C. Hoa, (2010) A dye-sensitized photo-supercapacitor based on PProDOT-Et2 thick films, *J Power Sources.* 195 6232–6238.

41. Z. Yang, L. Li, Y. Luo, R. He, L. Qiu, H. Lina, & H. Peng, (2012) An integrated device for both photoelectric conversion and energy storage based on free-standing and aligned carbon nanotube film, *J. Mater. Chem. A.* 1(3) 954–958.

42. R. Pourebrahim, S. Tohidi, & H. Khounjahan, (2021) Chapter 3 - Overview of energy storage systems for wind power integration, in: B. Mohammadi-Ivatloo, A. Mohammadpour Shotorbani, A. Anvari-Moghaddam (Eds.), *Energy Storage Energy Mark*, Academic Press, 41–64.

43. S.K. Bin Wu, Y. Lang, & N. Zargari, (2011) *Power Conversion and Control of Wind Energy Systems*, Wiley-IEEE Press.

44. A. Siria, M.-L. Bocquet, & L. Bocquet, (2017) New avenues for the large-scale harvesting of blue energy, *Nat. Rev. Chem.* 1 91.

45. G. Ramasamy, P.K. Rajkumar, & M. Narayanan, (2021) Generation of energy from salinity gradients using capacitive reverse electro dialysis: A review, *Environ. Sci. Pollut. Res.* 5 1–10.

46. B.E. Logan, & M. Elimelech, (2012) Membrane-based processes for sustainable power generation using water, *Nature.* 488 313–319.

47. E. Brauns, (2008) Towards a worldwide sustainable and simultaneous large-scale production of renewable energy and potable water through salinity gradient power by combining reversed electrodialysis and solar power?, *Desalination.* 219 312–323.

48. S. Loeb, (1975) Osmotic power plants, *Science.* 189 654.

49. M. Elimelech, & W.A. Phillip, (2011) The future of seawater desalination: Energy, technology, and the environment, *Science.* 333 712–717.

50. M. Olsson, G.L. Wick, & J.D. Isaacs, (1979) Salinity gradient power: Utilizing vapor pressure differences, *Science.* 206 452–454.

51. I. Ihsanullah, M.A. Atieh, M. Sajid, & M.K. Nazal, (2021) Desalination and environment: A critical analysis of impacts, mitigation strategies, and greener desalination technologies, *Sci. Total Environ.* 780 146585.

52. P. Długołęcki, K. Nymeijer, S. Metz, & M. Wessling, (2008) Current status of ion exchange membranes for power generation from salinity gradients, *J. Memb. Sci.* 319 214–222.

53. J.W. Post, H.V.M. Hamelers, & C.J.N. Buisman, (2008) Energy recovery from controlled mixing salt and fresh water with a reverse electrodialysis system, *Environ. Sci. Technol.* 42 5785–5790.

54. F. La Mantia, M. Pasta, H.D. Deshazer, B.E. Logan, & Y. Cui, (2011) Batteries for efficient energy extraction from a water salinity difference, *Nano Lett.* 11 1810–1813.

55. Z. Jia, B. Wang, S. Song, & Y. Fan, (2014) Blue energy: Current technologies for sustainable power generation from water salinity gradient, *Renew. Sustain. Energy Rev.* 31 91–100.

56. D. Brogioli, R. Zhao, & P.M. Biesheuvel, (2011) A prototype cell for extracting energy from a water salinity difference by means of double layer expansion in nanoporous carbon electrodes, *Energy Environ. Sci.* 4 772–777.

57. P. Sharma, & T.S. Bhatti, (2010) A review on electrochemical double-layer capacitors, *Energy Convers. Manag.* 51 2901–2912.

58. D. Brogioli, (2009) Extracting renewable energy from a salinity difference using a capacitor, *Phys. Rev. Lett.* 103 58501.

59. V. Khomenko, E. Raymundo-Piñero, & F. Béguin, (2010) A new type of high energy asymmetric capacitor with nanoporous carbon electrodes in aqueous electrolyte, *J. Power Sources.* 195 4234–4241.

60. N.Y. Yip, D. Brogioli, H.V.M. Hamelers, & K. Nijmeijer, (2016) Salinity gradients for sustainable energy: Primer, progress, and prospects, *Environ. Sci. Technol.* 50 12072–12094.

61. N.-L. Wu, S.-Y. Wang, C.-Y. Han, D.-S. Wu, & L.-R. Shiue, (2003) Electrochemical capacitor of magnetite in aqueous electrolytes, *J. Power Sources.* 113 173–178.

62. K.K. Patel, T. Singhal, V. Pandey, T.P. Sumangala, & M.S. Sreekanth, (2021) Evolution and recent developments of high performance electrode material for supercapacitors: A review, *J. Energy Storage.* 44 103366.

63. S. Kumar, G. Saeed, L. Zhu, K.N. Hui, N.H. Kim, & J.H. Lee, (2021) 0D to 3D carbon-based networks combined with pseudocapacitive electrode material for high energy density supercapacitor: A review, *Chem. Eng. J.* 403 126352.

64. R.B. Choudhary, S. Ansari, & M. Majumder, (2021) Recent advances on redox active composites of metal-organic framework and conducting polymers as pseudocapacitor electrode material, *Renew. Sustain. Energy Rev.* 145 110854.

65. D.P. Chatterjee, & A.K. Nandi, (2021) A review on the recent advances in hybrid supercapacitors, *J. Mater. Chem. A.* 9 15880–15918.

66. M. Pasta, C.D. Wessells, Y. Cui, & F. La Mantia, (2012) A desalination battery, *Nano Lett.* 12 839–843.

12 Market Trends, Innovations and Challenges

Anjali Paravannoor

CONTENTS

DOI: 10.1201/9781003258384-12

12.1 INTRODUCTION

The application of various supercapacitor (SC)-based energy storage systems is emerging as an easy and viable solution to present-day power distribution system applications, owing to the advancements in the field of energy storage technologies and tailored devices as well as the flexibility in design and configuration it can offer [1–5]. At the same time, there are challenges to be addressed in the power sector as well as the technology solutions, especially since the market itself has had a transition from regulated pattern to a much deregulated one recently and is centralized to localized systems [6–8]. The major constraints include maintaining a real-time flow of electricity while also addressing a range of concerns from faults in the distribution systems and inconsistencies in the renewable energy sources, to environmental sustainability and operational flexibility [9, 10]. This chapter deals with the various market considerations, innovative technologies in the field and also discusses the challenges to be addressed.

12.2 MARKET CONSIDERATIONS

The SC market has comfortably positioned itself such as to fill the gap between the capacitor market and battery market, two large industries, extended across the globe, which would obviously oblige the battery manufacturers and capacitor manufacturers to enter the SC market [11–13]. SCs offer a higher energy density compared to conventional capacitors and a higher power density compared to batteries. In the application perspective, SCs have marked their presence in a broad spectrum of end uses ranging from consumer electronics and communication systems to renewable energy and the automobile industry. In 2020, the global SC market was valued at USD887 million and it is expected that by the end of 2026, it could reach about USD1870 million at a compound annual growth rate (CAGR) of 13.5% during this period [13].

Considering conventional energy storage solutions, the introduction of SC technology can either augment existing battery systems or it can replace traditional batteries and/or capacitors. It can also be positioned and proposed in independent industrial and automobile applications [10, 14, 15].

12.2.1 DEMAND CREATION IN THE EXISTING MARKET

12.2.1.1 Hybrid Energy Storage Systems

One of the early applications of first-generation SCs includes the augmentation of the battery market. The SCs are used here for load-levelling so as to take up the peak power requirement by assisting banks of lead-acid, nickel-metal hydride or lithium-ion (Li-ion) batteries [16, 17]. Such hybrid systems are gaining attention in various applications like electric vehicles, electric grids and electric construction machinery. The inclusion of SCs in the system is highly advantageous owing to their higher rate capability, specific power values, low internal resistance and long cycle life which would enhance the life expectancy of the system to a great extent.

12.2.1.2   Replacement of Traditional Battery Systems

By far the largest market for SCs is in consumer electronics like video players and televisions wherein the coin cell double-layer SCs based on carbon are employed for clock functions, replacing Li-ion batteries. Such SCs are comparable to Li-ion batteries in price while ensuring much longer life expectancy and are discharged at a slower rate, enabling clock functions to continue even during a power outage. However, the replacement of batteries in broader application areas still remains a challenge and efforts are underway to improve the energy density of SCs and hence extend their application areas [18].

12.2.1.3   Replacement of Traditional Capacitor Systems

The high energy densities of supercapacitors as compared to conventional aluminium electrolytic capacitors (AECs) make them a potential replacement for AECs which in turn would reduce the size of electric circuits. Research is already underway into this for different applications like starter motors, base line stiffeners and AC line filters [19–21]. For instance, AECs are used as AC line filters in most of the line-powered electronics to attenuate the leftover AC ripples on DC voltage busses, while they are usually the largest component in the circuit and replacing the AECs with different carbon-based electric double-layer capacitors (EDLCs) has shown promising results [22].

12.2.2   MARKET CONSIDERATIONS SPECIFIC TO END USERS

The ever-increasing demand for a sustainable energy future while also addressing environmental and safety concerns has increased the pace towards novel energy generation and storage systems [2, 5, 23]. This is mainly achieved by a progressive replacement of fossil fuels with renewable energy sources like the sun, tide and wind [24–26]. However, efficient energy storage systems are required to surmount the intermittent and unpredictable nature of these renewable energy sources and in most of the cases batteries are the technology of choice [27, 28]. However, the safety concerns, low power density and low cyclic stability of battery systems make SCs one of the highly sought technologies addressing these issues, for energy generation as well as other applications like automobile electronics, power grids, consumer electronics and other industrial electronics [1–5].

12.2.2.1   Automobile Electronics

Road transportation-related industries including electric vehicles (EVs) and hybrid electric vehicles (HEVs) are one of the fastest expanding markets, replacing the internal combustion engine [29–31]. It is expected that over 50 HEV models will be on the road by 2030, with an outstanding 300% growth rate which in turn would last at least for a half decade [32]. Apart from being a green technology, other factors like the emergence of driverless technology as well as the evolution of grid infrastructure and other public conveyance systems such as light rail also act as driving forces for moving towards EVs and HEVs [33, 34]. Development of a dual power source for traditional automobiles is another highly sought technology owing to the new electronic systems involved, like navigation units, active suspension and radar circuits. For example, load-levelling by the inclusion of a SC would be a very promising approach. In a broader context there are several applications also related to the automobile industry where SCs are placed including actuators, electric power steering and gasoline engine starters. For instance, with the introduction of electric power steering which would require a SC system of 30–50 F and 16 V, about 1.5 gallons of gas can be saved per mile [35]. Similarly, gasoline engine starting or EV starting could be aided by SC systems in multiple domains like load-levelling of the batteries, aided acceleration, retrieval of braking energy, etc. [36]. SCs rated between 10 and 15 V could also be used as audio base line stiffeners in the audio systems of vehicles for levelling the load from an enormous base line so that other electrical systems are not affected. The application of the SCs would also need the right choice of design and configuration as apart from the first generation EDLCs, the market for pseudocapacitors and hybrid capacitors is expanding rapidly as perfect candidates for EV and HEV applications owing to their high energy density maintaining power density.

12.2.2.2   Industrial Electronics

There are various industrial applications like locomotive engines and marine engines that would incorporate SCs of wide voltage and capacitance ratings ranging from 10 to 300 V and 1 to 250 F respectively [37]. Few of the applications include the engine circuit. In most cases, a bank of AECs is included as these industrial engine circuits range from 150 to 300 V with a capacitance requirement of 0.1 to 2 F. SCs are being widely projected in this context as a replacement to aluminium capacitors (AECs) [19, 20]. There are other applications also like switches and actuators, uninterrupted power supplies, etc. [38, 39].

12.2.2.3   Consumer Electronics

Portable devices and consumer electronics hold one of the largest shares in the SC market. They are becoming smaller and thinner and require high power capabilities owing to the multiple functional trends being developed at the device level that require variable power profiles [40]. Conventional Li-based battery technologies are becoming less capable of providing the high power density needed and SCs are emerging as a potential solution in this regard.

12.3   INNOVATIVE TECHNOLOGIES AND FUTURE PERSPECTIVES

12.3.1   NOVEL MATERIALS

The innovative ideas in the development of SCs are mostly centred around the development of novel electrode materials as it is the most determining factor in any SC system. To augment the overall performance of SCS, researchers have put great effort into finding novel electrode materials whose properties play a vital role in enhancing the performance of SCs [41]. Efforts are being put into finding better alternatives for conventional electrode materials based on carbon, transition metal oxide and conducting polymer.

In recent years, metal organic frameworks (MOFs) are among one of the trending electrode materials for SCs, and this could be attributed to their fascinating properties including diverse structure, tuneable porosity, flexibility in making composite structures with faradaic materials and controllable chemical composition. In one of the early attempts in the field of MOFs based on Ni as the redox component reported a specific capacitance of 634 Fg^{-1} [42]. In order to address the low electrical conductivity of typical MOFs, $Ni_3(2,3,6,7,10,11$-hexaiminotriphenylene)$_2$ ($Ni_3(HITP)_2$) with high electronic conductivity (45,000 S m^1) was presented as a free-standing EDLC electrode and a capacitance value of 111 Fg^{-1} was attained with triangular traces [43]. Efforts are underway in further improving the performance of MOF-based SCs by further modifications [44].

Covalent organic frameworks (COFs) constitute yet another class of interesting electrode material which are highly porous and crystalline polymers wherein organic building blocks are carefully arranged into an ordered 2D or 3D array with atomic level precision. Strong covalent bonds like B-O, C-N, B-O-Si are formed through highly predictable organic reactions which provide flexibility in the molecular design, porosity as well as surface area. The incorporation of redox active molecules could also overcome the low hydrolytic and oxidative stability of the COFs. For example, redox-active 2,6-diaminoanthraquinone (DAAQ) was included in the COFs by beta ketoenamines to impart explicit redox reactions [45]. Similarly, a TaPa-Py COF is a pyridine-containing COF which shows well defined redox processes in 1 M H_2SO_4 electrolyte [46].

Considering their high electronic conductivity, metal nitrides of V, Ti, Ga, Mo, etc., are also considered as promising electrode materials [47]. For instance, VN possesses multiple advantages in terms of conductivity, theoretical capacity (4200 Fg^{-1}), high overpotential for hydrogen evolution and multiple oxidation states of vanadium enabling fast redox processes. When combined with different acidic, neutral and alkaline aqueous electrolytes of H_2SO_4, $NaNO_3$ and KOH, specific capacitance of 114, 45.7 and 273 F g^{-1} respectively were obtained [48]. G-Mo_2N and NbN-based electrodes could exhibit 111 and 73.5 F g^{-1} but with a narrow potential window. Areal capacitance value up to 23.11 mF cm^{-2} was observed for GaN in acidic electrolyte which could be further improved with the presence of oxynitrides.

2D nanostructures are one of the most fascinating classes of electrode material owing to their high surface area and flexibility. Black phosphorous is one of the newest members in this class and has been attracting the attention of researchers as it possesses a very unique structure and chemical as well as physical properties and efficient electrochemical performance as a SC electrode [49]. In one of the recent studies, Hao et al. designed a flexible solid-state SC with exfoliated black phosphorous flakes as the active electrode material [50]. Volumetric capacitance up to 17.78 Fcm^{-3} was attained with a capacitance fade of less than 30% after 30,000 cycles. More interestingly, oxygen functional groups are formed on the surface of black phosphorous in the presence of water which in turn can serve as active sites for intercalation/deintercalation of protons enhancing the capacitance values. Novel electrolyte formulations are also being developed to further improve the storage efficiency, rate capability and cyclic stability of various SC systems.

12.3.2 Technological Developments

12.3.2.1 Flexible Supercapacitors

With the drastic advances in the field of portable devices and wearable electronics, flexible energy storage technologies have become a hotspot for research [51, 52]. While the unbending nature of conventional storage devices restrict the shape and size of such consumer electronic devices, flexible SCs with high electrochemical performance can be considered to be one of the major developmental areas in the field of energy storage research. This is particularly important when the wearables include e-textiles while the inclusion of an externally powered storage device like a battery is not a feasible approach [53]. All the constituents of the system including cathodes, anodes, fluid collectors and packaging units should be flexible; however, in the functional aspect, development of a suitably flexible electrode is the major challenge and research is being focused on this area. Flexible SCs were primarily based on a flexible and conducting carbon network serving as current collector as well as active electrode material [54].

12.3.2.2 Micro Supercapacitors (MSCs)

Micro supercapacitors (MSCs) have drawn considerable attention in the recent years, especially due to the advancement in smart microelectronics. MSCs are miniaturized SCs with their size ranging from few micrometres to centimetres and could be easily integrated with diverse functional devices like MEMS, wireless networks and potable and wearable electronics [55]. Owing to the better power density provided by MSCs compared to microbatteries, that too in a smaller space, makes them highly suitable in this context. Moreover, MSCs have very long cyclic stability which gives them higher practicability, especially for certain applications [56]. For instance, when used in powered biological implants like biochips the longer lifespan of MSCs addresses the most important challenge of frequent replacement associated with microbatteries. Along with the economic and environmental benefits this would also iron out the difficulties of invasive methods like a surgery associated with the replacement [57].

Looking deeper into the technology, MSCs can be of three different configurations: Traditional sandwich structures, inter-digital type and fibrous type; however, the pros and cons of each of these configurations depends upon the specific applications and requirements. For instance, the interdigitated configuration has better efficiency and compatibility in on-chip applications owing to the facile integration on plane, better ion diffusion efficiency, etc., while fibre-type MSCs are more beneficial in wearable electronics due to structural advantages [58].

Research is underway in developing various components of MSCs like electrodes, electrolyte separators and current collectors to improve efficiency; nevertheless, electrodes are the central constituent that largely affects the overall performance of the system and tremendous progress has happened over recent years in EDLCs as well as pseudocapacitor electrodes. Porous nano/micro architectures are being developed; for example, to improve the double-layer capacitance of MSC systems wherein there will be an increase in number of active sites. Li et al. fabricated a MSC system that could deliver a stack power of 232.8 W cm^{-3} at a scan rate as high as 1000 mVs^{-1} and a MOF-CVD-based method was followed for the synthesis of porous carbon coated interdigitated gold electrodes [59]. Similarly, He et al. developed another microfabication approach for influencing the arrangement of nanostructured building units by which multi-walled carbon nanotubes (MWCNTs) are arranged on the surface of pyrolyzed carbon microelectrodes to obtain areal and volumetric capacitance of 4.8 mFcm^{-2} and 32 Fcm^{-3} respectively [60]. Such ordered arrangements would impart an increased active area on the electrode overlay by avoiding overlapping of individual units (CNTs in this case) which would increase the number of active sites as well as the contact area of electrolyte ions, which in turn would improve the energy density of the system. Moreover, there has been a multitude of works based on graphene and its modification via doping or composite approaches to be used as MSC electrodes [61].

Pseudo MSCs are getting much attention owing to their higher energy performance compared to EDLC MSCs while research is being focused on improving the power density and cyclic stability of such systems. For instance, Lee et al. developed a one-step laser fabrication of self-generated nanoporous metal/oxide electrode for MSCs [62]. Hetero metal oxides (manganese and iron oxides) were used as the active materials, whereas porous structures of nanostructured silver self-generated on a polymer film acted as efficient current collectors. A volumetric energy density of 16.3 mWhcm^{-3} was achieved at a power density as high as 3.54 Wcm^{-3}. Moreover, an asymmetric configuration was used which could increase the operating voltage. In yet another work by Kurra et al., poly(3,4-ethylenedioxythiophene) (PEDOT)-based electrodes were synthesized following photolithography combined with electrochemical deposition and ultra-high rate capability was achieved with an areal capacitance of 9 mFcm^{-2} [63]. Remarkable performance was observed with a tuneable frequency response applicable for AC-line filtering applications.

12.3.2.3 Hybrid Capacitors

Hybrid SCs, which are the assemblage of EDLC and faradaic capacitor exhibit enhanced characteristics compared to the components combined. This is considered to be one of the most efficient ways to improve the energy density of SCs. Metal-ion capacitors are the most studied and developed area with lithium-/sodium-/potassium-ion capacitors exhibiting a high energy as well as power performance with ultra-long life span. For example, a sodium-ion hybrid capacitor was developed by Wang et al. based on carbon with a high energy density of 168 Wh kg^{-1} and the highest power density going up to 2432 W kg^{-1} (@ 98 Wh kg^{-1}) and a capacitance retention of 85% at the end of 1200 consecutive cycles [64]. Storage mechanisms using polyvalent ions are one of the very recent developments in this regard involving zinc-, magnesium-, calcium- and aluminium-based batteries to compete with the Li-ion battery technology and the trend has been extended to polyvalent ion hybrid capacitors with zinc-ion capacitors (ZIC) taking the lead. Efforts are being made to improve the cyclic stability while preventing the disadvantageous like dendrite growth in ZICs. Yin et al. designed a ZIC by incorporating hydrogen and oxygen redox reactions in aqueous ZICs to enhance storage capability and a high capacitance of up to 340 Fg^{-1} could be obtained along with a power density of 48.8 kW kg^{-1} at an energy density of 104.8 Wh kg^{-1}. Interestingly, the cyclic stability was also very high with a cycle life as long as 30,000 cycles with less than 1% capacitance fade [65].

Apart from the metal ion capacitors, hybrid capacitors can also be asymmetric systems with EDLC/metal oxide configurations. A few such examples include carbon-based materials combined with metal oxide or conducting polymer-based pseudocapacitor electrodes where an activated carbon-lead oxide system is considered to be the best configuration. Other most common designs include Co- or Ni-based oxide or hydroxide electrodes being the positive electrode and a C-based negative electrode like activated carbon, rGO, carbon aerogel or carbon nanofibre. The Ni(OH)$_2$-rGO system devised by Mohammed et al. could yield a specific energy close to 64 Wh kg^{-1} with a corresponding power density of 1 kW kg^{-1} [66]. Similarly, an interfacial polymerized PANI-PMo$_{12}$ hybrid material could deliver a specific capacitance ranging from 68.8 to 172.4 Fg^{-1} depending upon the synthesis conditions [67].

Hybrid SCs have found various applications ranging from public sector and EV/HEV applications to military and biomedical applications. Initially, the public sector usage of SCs wes limited to actuator and memory backup applications. Then different SC systems applicable to remote controls, flashes of digital cameras, security systems, etc., evolved. Maxwell Technologies, one of the leading SC manufacturers replaced AAA batteries in remote controls with fast charging SCs and it is projected that they could even exceed the lifetime of the remote. Similarly, when used in portable speakers the in-built SC assembly could provide a play back time of six hours after five minutes charging [68].

The integration of new and renewable energy sources like wind, tide and sun with SCs is another leading application field of SCs. Nippon-Chemi-Con were providing eco-friendly street lighting in Japan by combining SCs with solar cells from as early as 2010 [75]. As discussed earlier, the most fascinating application of SCs is in the automobile sector. Hybrid SCs are used for power bursts in Toyota's Yaris Hybrid-R, for example. Aowei Technology Co., Ltd. (China) pioneered e-buses and trolley buses with quick charging ability [69]. The trolley buses use Ni(OH)$_2$-AC hybrid configuration and with a charging time as low as 90 seconds, they can achieve an average distance 7.9 km, with the speed being about 44 kmh^{-1} [70]. Hybrid SCs are also widely used in regenerative braking applications. Different defence applications like sensors, navigators, radar systems, torpedoes and different communication tools are all run by batteries and most of them could be easily replaced with hybrid SCs of optimal designs and configurations. They have also found applications in computers, memory chips, biochips, ventilator backups and electricity grids.

12.3.2.4   Piezoelectric SCs

Piezoelectric SCs are self-charging devices that are capable of directly converting mechanical energy to electrical energy and have been rapidly gaining interest owing to the advancements in wearable and stretchable electronics. Making use of their outstanding flexibility and mechanical stability, functionalized carbon clothes are being extensively used in this regard. In one of the recent works, carbon clothes are used as both electrodes along with polarized PVDF films and H$_2$SO$_4$/PVA gel electrolyte. A specific capacitance as high as 357 Fm^{-2} was attained at a current density of 8 Am^{-2} with a good rate capability. Under constant compressive force of 4.5 Hz frequency, the voltage could be increased to 100 mA within 40 seconds [71]. Utilization of redox material is also being thoroughly investigated. One such configuration includes a symmetric design of MnO_2 as both the electrodes and a PVDF-ZnO separator become piezoelectric material. The voltage of the device was found to increase from 35 to 145 mV in 300 seconds upon the compressive stress caused by constant palm impact [72].

12.3.2.5 Shape Memory SCs

Shape memory SCs constitute yet another class of novel storage device wherein specific shape memory materials are incorporated so as to regain the shape and/or size of the device followed by irreversible distortions that may happen in practical conditions. Shape memory materials can be based on alloys (SMA) or polymer materials (SMP). After deformation the SMA-based materials recovered when heated to specific temperatures, by removing all the plastic deformations. Making use of the most studied shape, SMA, NiTi, as the negative electrode after coating with graphene and a thin MnO_2/Ni film as positive electrode, a specific capacitance of 53.8 Fg^{-1} was attained and after heating to ambient temperature, the deformed device could recover its original shape within 550 seconds [73]. SMPs exhibit much more interesting properties as they memorise multiple shapes and can be recovered after deformation by subjecting to various stimuli, like temperature, electric current, magnetic field, light, etc. For instance, aligned CNT sheets were wrapped in a polyurethane based SMP and used in wire shaped MSCs and at the end of 500 cycles of deformation and recovery (50% strain), electrochemical performance remained unchanged indicating outstanding stability [74]. However, the maximum specific capacitance was limited to about 42 $mFcm^2$. Hence researchers are putting in effort to improve the storage performance by the incorporation on faradaic materials into such systems.

12.3.2.6 Transparent Supercapacitors and Others

The rapid growth in the field of optically transparent portable and wearable intelligent devices like phones, tablets, displays, panels and other touchable devices requires fast progression in the development of matching power sources. Transparent and flexible SCs are the most suitable choice owing to their good optical transmittance together with flexibility and efficient electrochemical performance. Various functional nanostructures in the pristine and composite form including graphene, CNTs, conducting polymers and metal oxide thin films are a hotspot research area to be used in transparent supercapacitors (TSCs), in their thin film form on suitable transparent current collectors. Owing to their efficient electrochemical performance, tuneable morphology and porosity, facile fabrication techniques and high electrical conductivity, carbon-based electrodes are the most sought out electrode materials for TSCs in spite of their transparency being limited. Hence, research is underway to improve the transparency of such electrodes. A CNT-based electrode was developed by King et al. with ferrocene as the source material and the film exhibited 90% transparency and a sheet resistance of 41 Ωs/q [75]. Owing to their better energy profile metal oxide-based thin film electrodes are also being widely investigated. Self-supporting ultra-thin micro-nanometal grid/nano-manganese oxide composite electrode materials were developed by Xu et al. along with a water base gel electrolyte, PVA/LiCl [76]. An ultra-thin electrode could be fabricated with the device thickness being <20 μm with excellent storage efficiency and cyclic stability and 80% transparency. The device could also exhibit exceptional mechanical flexibility and could even be kneaded into a ball. In spite of the considerable number of efforts on the topic, development of efficient TSCs with good chance of scalability still needs better research focus and constant effort that would also be beneficial in terms of safety and environmental friendliness.

There are more innovative technologies based on SCs being proliferated with the advent of intelligent electronics as it demands the intelligentization and better controllability of energy storage devices. They are expected to be multifunctional and customizable making them more user-friendly providing personalized interaction with electronic devices. For instance, a water-activated storage device was assembled by Yin et al. from biodegradable materials [77]. In another interesting work, the concept of edible SCs was projected by Wang et al., by fabricating capsule-sized fully functional SC units [78].

Solar energy is the most environmentally benign and renewable energy source; integrated devices of SCs and solar cells is another fast-growing area in SC research. This is especially significant considering the unpredictable nature of the source. Hence different configurations of photo self-charging SCs are being evolved. For instance, Yu et al. developed a fibre photo-SC where a fibre shaped DSSC in integrated with a fibre shaped MSC [79]. The two constituent electrodes, Ti wire and CNTs were twisted together, and the overall photoelectric conversion and storage efficiency was found to be about 1.5%.

12.3.3 Developments in the Application Scenario

12.3.3.1 Social Demands

While the rapid developments in the electronics industry demand better and multifunctional storage devices, research and people across the globe are working to make the solution cleaner and more environmentally benign. Hence the advancements in SC technology are a social need and have a broad market prospect. The initial target of the US Department of Energy while entering into SC research was to achieve an energy density of 5 Wh kg^{-1} in 1992 which was already reached. A specific energy of 6 Wh kg^{-1} with a power density of 1.5 kW kg^{-1} was then attained which could easily meet the requirements of electrochemical cells as well as fuel cell EVs. The target now is to reach an energy density of 20 Wh kg^{-1} that would be ideal for hybrid cars which is primarily aiming at social advancement [80].

12.3.3.2 Scalability

Considering the application prospects, scalability is another most important factor to be considered and along with the improvements in the performance of the product, it demands better control over the production cost. This would require a careful choice of raw materials, optimizing easier methods of fabrication and production equipment. For instance, in order to lower the overall cost, low-price raw materials can be combined with high-cost raw materials so as to maintain product quality. Hence, future research should also consider the social requirements along with the technological aspects.

12.4 CHALLENGES ASSOCIATED WITH DEVELOPMENT OF SUPERCAPACITORS

12.4.1 TECHNICAL CHALLENGES

The most crucial factor associated with the development of SCs is to optimize the energy and power density with long cycle life and stability, meeting the demands for next-generation electronic applications. In spite of the higher energy profile of SCs compared to the traditional parallel plate capacitors, it is not even comparable with batteries and fuel cells. Hence the various facets of SCs are all aimed at a single point: Improving the energy density without compromising the power density and cyclic stability. The performance of SCs can be attributed to various key parameters such as potential operating window, equivalent series resistance (ESR), full cell voltage and time constant. The fabrication and commercialization of SCs still needs a better understanding of these parameters while investigating the structure of electrodes, electrode/electrolyte interaction, and the mechanism involved. Commonly, the overall performance of a SC mainly depends on various key components including electrode materials, electrolytes and device configuration. Researchers also give attention to the choice of cheaper materials and methods so that it is economical when incorporated in respective electronic designs.

Apart from these challenges, in general there are specific research areas to be developed to match specific applications. For example, a better modelling of the system is needed so that many of the non-ideal parameters, which are insignificant otherwise, have to be taken into account. For instance, the resonance caused by signal and filter would have significant impact in certain power supply applications like satellites and spaceships. Similarly, the impact of various factors on the stability of the system should also be taken care of while establishing a model [81].

12.4.1.1 Electrode Materials

In EDLCs, carbonaceous electrode materials are mainly utilized due to their unique properties and none among the various carbon-based electrode materials can be considered as impeccable. Activated carbon electrodes are the most widely used carbon-based electrode material due to their tuneable porosity and high surface area; however, they suffer from lower electrochemical performance and electrical conductivity. Various strategies are utilized to enhance the electrochemical performance of activated carbon like using ultrasonic radiation modifying the Fermi level position, doping sulphur and oxygen functionalities, surface modification of activated carbon via oxidation process, fabricating composite of the electrode with other carbon materials or by inserting polymers into the carbon substrate [82]. CNT electrodes exhibit comparatively low ESR and higher power than AC electrodes owing to their specific mesoporous internal network which helps in easy and faster diffusion of electrolyte ions [83]. Though they act as potential material for high-power EDLC electrodes, one of the major demerits associated with CNTs is their low specific surface area (<500 m^2 g^{-1}). The CNT-based electrode exhibits specific capacitance less than 200 F g^{-1} and thus one of the commonly adopted methods to improve specific capacitance of CNTs via addition of functionalities into the CNT matrix. Graphene-based electrodes address most of the concerns raised by other carbon-based electrodes and they exhibit unique properties like novel morphological structure, superior mechanical properties, good electrical conductivity, high carrier mobility, large specific surface area, etc. Graphene films have also been employed as stretchable electrodes [84]. Nonetheless, unfortunately, agglomeration of graphene nanosheets by virtue of a strong van der Waals interaction limits the direct access of graphene surface which in turn results in the increase of ionic resistance in the electrode [85]. Another major drawback with the commercialization of graphene electrodes is the difficulty in reliable production of high-quality graphene from a scalable and cost-effective method. Hence the development of suitable carbon electrodes which do not compromise performance or cost still remains a challenge.

Transition metal oxides (TMOs), conducting polymers and other redox-active SC electrodes exhibit higher specific capacitance (100–3000 F g^{-1}) and higher energy density compared to carbon electrodes [86]. One of the most ideal pseudocapacitive electrode materials is RuO_2 but its higher price and toxicity limits its application. RuO_2 possesses high theoretical capacitance (2000 F g^{-1}), high chemical and thermal stability, good electrical conductivity and a large voltage window. In order to reduce the cost of RuO_2 electrodes various composites of RuO_2 are synthesized, such as carbon material-RuO_2 electrode, metal oxide-RuO_2, metal sulphide RuO_2, etc., and other metal oxide and metal chalcogenide systems are also being investigated widely [87]. In spite of the advantages in terms of storage efficiency and energy density, the prime causes that make metal oxides less attractive is that they possess a strong bonding force with the electrolytes as it results in poor regeneration. They also suffer from low rate capability and also capacitance deterioration on cycling which reduces the lifespan. Moreover, synthesis of effective metal oxide nanostructure is challenging and costly. Thus, researchers are focusing on the study to improve the performance of transition metal oxides and also to design new materials to address these concerns. MXenes are another class of novel electrode materials, fascinating the researchers as they possess outstanding mechanical, electrical and electrochemical properties [88]. Despite the unique properties of MXenes, there are several drawbacks associated with them in the practical scenario. For instance, it is very difficult to maintain a balance between the electrochemical performance and mechanical properties of MXenes when they are used in flexible SCs. Another major concern is the tacking of individual layers which impart a significant hindrance to ionic diffusion in the vertical direction which in turn would affect the rate performance of electrodes [89]. The tendency of MXenes to oxidize is another major issue which would be further facilitated in the presence of defects on the surface or at the edge of the flakes and the crystalline MXene undergoes degradation.

12.4.1.2 Electrolytes

The formulation of an ideal electrolyte is challenging as umpteen requirements are to be considered like a wide potential window, chemical stability, wide operating temperature, low volatility, inertness when in contact with current collectors, packaging elements and separators, low flammability and cost. An electrolyte to meet all these requirements is hard as each formulation will have its own pros and cons. Most commonly used electrolytes are based on aqueous, organic or ionic liquids and they are further categorized into a myriads compositions. For instance, aqueous electrolytes can be acidic, alkaline or neutral and each of them behave differently [90]. When combined with acidic electrolytes, surface-quinone functionalities produce pseudocapacitive effects involving protons while it behaves as a pure EDLC electrode in alkaline electrolyte. While the redox performance is helpful in enhancing the energy density, it affects the power density [91]. Some of the best-performing neutral electrolytes like hexaflourosilicic acid or tetraflouroboric acid possess serious safety concerns [92]. The type and concentration of electrolytes would also affect the ESR value, specific capacitance and gas evolution reactions. When used in higher concentration, alkaline electrolytes can cause corrosion of the active electrode overlay in spite of the better ionic conductivity. Widening the voltage window and avoiding the unwanted side reactions the two major aims concerned with aqueous electrolytes and neutral electrolytes can solve these issues while also reducing electrode dissolution. However, the low energy density of the neutral electrolytes remains a challenge.

Even though aqueous electrolytes are being widely investigated in the research and academic aspect, the SC market is dominated by organic electrolytes as their potential window is significantly high (2.5–2.8 V) [90, 93–95]. However, they do suffer from many disadvantageous like lower specific capacitance as well as conductivity and higher volatility and flammability, raising safety concerns. Furthermore, they have higher cost and require complicated procedures for purification and assembly and a controlled environment which would also add to the cost. Moreover, the electrode surface area is usually improved by the introduction of micropores, which is not practical with organic electrolytes as larger ions will leave the pores unutilized. The higher voltage window, while improving the energy and power density, could also be detrimental while using faradaic electrodes as the metal oxides may get oxidized which in turn would affect the performance and stability. The presence of trace water, if any, can also cause self-discharge.

Ionic liquid-based electrolytes can also be used which are also capable of providing a higher potential window along with other potential merits like thermal and chemical stability, low volatility and inflammability [96, 97]. They can also be tuned with the limitless available combinations of cations and anions. Even with these distinct benefits, they suffer from the higher viscosity and lower conductivity causing higher ESR and also lead to lower specific capacitance values. Solid-state electrolytes are being widely investigated as a leakage-free material while they also possess advantages in terms of a facile assembly process. However, most of the solid electrolytes developed so far are polymer-based materials and they largely suffer from the lower conductivity and rate capability [98, 99]. Inorganic solid materials are considered to be better alternatives in this regard, but the available literature on this area is highly limited.

12.4.2 CHALLENGES IN THE APPLICATION PERSPECTIVE

Considering the pace at which the SC industry is growing and SCs, being a novel storage technology, the developments in the market should be carefully monitored and practical standards should be set. Various standards are to be considered, ranging from a naming method and safety requirements to material specifications and electrical performance test details. Non-technical details like transportation and disposal requirements should also be specified. While doing these, efforts are needed to make the whole process green and cost effective especially in disposing of the various components and recycling which are also a requirement of society. Similarly, in an application point of view, depending upon the specific requirement, different parallel and series connections would be necessary and hence consistency should be maintained in the specification of individual capacitors in the connection.

12.5 SUMMARY AND OUTLOOK

In the domain of energy storage technology, SCs are emerging as one of the most potential candidates as a replacement to traditional batteries and/or to augment existing battery technology. With varying design, configuration and specifications they have found numerous applications and end users ranging from automobiles and consumer electronics to renewable energy technology and defence applications. Considering the market requirements, drastic advancements in the field with the development of novel materials, processes and designs are also underway. However, the successful scaling and commercialization of the technology still faces several technological as well as societal challenges, and the energy research of the future will surely be centred around addressing these issues.

REFERENCES

1. Raghavendra, K. V., Vinoth, R., Kamran, Z., Chandu, V. V., Muralee, G., Sambasivam, S., Madhusudana, R. K., Ihab, M. O., & Hee, J. K. (2020). An intuitive review of supercapacitors with recent progress and novel device applications. *Journal of Energy Storage*, 31(2020), 101652.
2. Wu, N., Xue, B., Duo Pan, B. D., Renbo, W., Nithesh, N., Rahul, R. P., & Zhanhu, G. (2021). Recent advances of asymmetric supercapacitors. *Advanced Materials Interfaces*, 8(1), 2001710.
3. Lv, Y., Shifei, H., & Yufeng, Z. (2019). NBF tridoped 3D hierarchical porous graphitized carbon derived from chitosan for high performance supercapacitors. *Science of Advanced Materials*, 11(3), 418–424.
4. Zhang, J., Mauricio, T., Chong, R. P., Rahul, M., Marc, M., Nikhil, K., Yern, S. K. et al. (2016). Carbon science in 2016: Status, challenges and perspectives, *Carbon*, 98(70), 708–732.

5. Simon, P., Taberna, P. L., & Béguin, F., (2013). *Electrical Double-Layer Capacitors and Carbons for EDLCs*, In Supercapacitors, Wiley, Chap. IV, pp. 131–165.

6. Pandolfo, T., Ruiz, V., Sivakkumar S., & Nerkar, J., General Properties of Electrochemical Capacitors, In *Supercapacitors*, Chap. II, pp. 69–109. Wiley.

7. Berrueta, A., Ursúa, A., San Martín, I., Eftekhari, A., & Sanchis, P. (2019). Supercapacitors: Electrical characteristics, modeling, applications, and future trends. *IEEE Access*, 7, 50869–50896.

8. Choi, N.S, Chen, Z., Freunberger, S. A., Ji, X., Sun, Y.-K., Amine, K., Yushin, G., Nazar, L. F., Cho, J., & Bruce, P. G. (2012). Challenges facing lithium batteries and electrical double-layer capacitors. *Angewandte Chemie International Edition*, 51, 9994–10024.

9. Wang, Y., Song, Y., & Xia, Y. (2016). Electrochemical capacitors: Mechanism, materials, systems, characterization and applications. *Chemical Society Reviews*, 45(21), 5925–5950.

10. Zuo, W., Li, R., Zhou, C., Li, Y., Xia, J., & Liu, J. (2017). Battery-supercapacitor hybrid devices: Recent progress and future prospects. *Advanced Science*, 4(7), 1600539.

11. Sumangala, T. P., Sreekanth, M. S., & Rahaman, A. (2021). Applications of Supercapacitors. *Handbook of Nanocomposite Supercapacitor Materials III: Selection*, pp. 367–393. Springer, Cham.

12. Lasrado, D., Ahankari, S., & Kar, K. K. (2021). Global trends in supercapacitors. *Handbook of Nanocomposite Supercapacitor Materials III: Selection*, 329–365.

13. Zhao, J., & A. F. Burke (2021). Review on supercapacitors: Technologies and performance evaluation. *Journal of Energy Chemistry*, 59, 276–291.

14. Sani, A., Siahaan, S., & Mubarakah, N. (2018, February). Supercapacitor performance evaluation in replacing battery based on charging and discharging current characteristics. In *IOP Conference Series: Materials Science and Engineering* (Vol. 309, No. 1, p. 012078). IOP Publishing.

15. Chotia, I., & Chowdhury, S. (2015, November). Battery storage and hybrid battery supercapacitor storage systems: A comparative critical review. In *2015 IEEE Innovative Smart Grid Technologies-Asia (ISGT ASIA)* (pp. 1–6). IEEE.

16. Liu, H., Wang, Z., Cheng, J., & Maly, D. (2008). Improvement on the cold cranking capacity of commercial vehicle by using supercapacitor and lead-acid battery hybrid. *IEEE Transactions on Vehicular Technology*, 58(3), 1097–1105.

17. Jiao, Y., Pei, J., Yan, C., Chen, D., Hu, Y., & Chen, G. (2016). Layered nickel metal–organic framework for high performance alkaline battery-supercapacitor hybrid devices. *Journal of Materials Chemistry A*, 4(34), 13344–13351.

18. Shukla, A. K., Banerjee, A., Ravikumar, M. K., & Jalajakshi, A. (2012). Electrochemical capacitors: Technical challenges and prognosis for future markets. *Electrochimica Acta*, 84, 165–173.

19. Saifulin, R., Pajchrowski, T., & Breido, I. (2021). A buffer power source based on a supercapacitor for starting an induction motor under load. *Energies*, 14(16), 4769.

20. Sheng, K., Sun, Y., Li, C., Yuan, W., & Shi, G. (2012). Ultrahigh-rate supercapacitors based on electrochemically reduced graphene oxide for ac line-filtering. *Scientific Reports*, 2(1), 1–5.

21. Joseph, J., Paravannoor, A., Nair, S. V., Han, Z. J., Ostrikov, K. K., & Balakrishnan, A. (2015). Supercapacitors based on camphor-derived meso/macroporous carbon sponge electrodes with ultrafast frequency response for ac line-filtering. *Journal of Materials Chemistry A*, 3(27), 14105–14108.

22. Han, Z. J., Huang, C., Meysami, S. S., Piche, D., Seo, D. H., Pineda, S., & Grobert, N. (2018). High-frequency supercapacitors based on doped carbon nanostructures. *Carbon*, 126, 305–312.

23. Chu, S., & Majumdar, A. (2012). Opportunities and challenges for a sustainable energy future. *Nature*, 488(7411), 294–303.
24. Hussain, A., Arif, S. M., & Aslam, M. (2017). Emerging renewable and sustainable energy technologies: State of the art. *Renewable and Sustainable Energy Reviews*, 71, 12–28.
25. Chowdhury, M. S., Rahman, K. S., Selvanathan, V., Nuthammachot, N., Suklueng, M., Mostafaeipour, A., & Techato, K. (2021). Current trends and prospects of tidal energy technology. *Environment, Development and Sustainability*, 23(6), 8179–8194.
26. Khare, V., Nema, S., & Baredar, P. (2016). Solar–wind hybrid renewable energy system: A review. *Renewable and Sustainable Energy Reviews*, 58, 23–33.
27. Notton, G., Nivet, M. L., Voyant, C., Paoli, C., Darras, C., Motte, F., & Fouilloy, A. (2018). Intermittent and stochastic character of renewable energy sources: Consequences, cost of intermittence and benefit of forecasting. *Renewable and Sustainable Energy Reviews*, 87, 96–105.
28. Liang, X. (2016). Emerging power quality challenges due to integration of renewable energy sources. *IEEE Transactions on Industry Applications*, 53(2), 855–866.
29. Chan, C. C. (2002). The state of the art of electric and hybrid vehicles. *Proceedings of the IEEE*, 90(2), 247–275.
30. Higueras-Castillo, E., Molinillo, S., Coca-Stefaniak, J. A., & Liébana-Cabanillas, F. (2019). Perceived value and customer adoption of electric and hybrid vehicles. *Sustainability*, 11(18), 4956.
31. Singh, K. V., Bansal, H. O., & Singh, D. (2019). A comprehensive review on hybrid electric vehicles: Architectures and components. *Journal of Modern Transportation*, 27(2), 77–107.
32. Hertzke, P., Müller, N., Schenk, S., & Wu, T. (2018). The global electric-vehicle market is amped up and on the rise. *McKinsey Cent. Futur. Mobil*, 1–8.
33. Harighi, T., Bayindir, R., Padmanaban, S., Mihet-Popa, L., & Hossain, E. (2018). An overview of energy scenarios, storage systems and the infrastructure for vehicle-to-grid technology. *Energies*, 11(8), 2174.
34. Cheng, L., Wang, W., Wei, S., Lin, H., & Jia, Z. (2018). An improved energy management strategy for hybrid energy storage system in light rail vehicles. *Energies*, 11(2), 423.
35. Szabo, A. Electrically powered steering using a supercapacitor power boost unit. In Conference: 4th IET International Conference on Power Electronics, Machines and Drives (PEMD 2008).
36. Passalacqua, M., Lanzarotto, D., Repetto, M., Vaccaro, L., Bonfiglio, A., & Marchesoni, M. (2019). Fuel economy and ems for a series hybrid vehicle based on supercapacitor storage. *IEEE Transactions on Power Electronics*, 34(10), 9966–9977.
37. Kozłowski, M. (2005). Application of supercapacitors to recuperate energy in diesel-electric locomotives. *Archives of Electrical Engineering*, 54(2), 205–224.
38. Saponara, S. (2016). An actuator control unit for safety-critical mechatronic applications with embedded energy storage backup. *Energies*, 9(3), 213.
39. Zhang, X., Xue, H., Xu, Y., Chen, H., & Tan, C. (2014). An investigation of an uninterruptible power supply (UPS) based on supercapacitor and liquid nitrogen hybridization system. *Energy Conversion and Management*, 85, 784–792.
40. Phoosomma, P., Kasayapanand, N., & Mungkung, N. (2019). Combination of supercapacitor and ac power source in storing and supplying energy for computer backup power. *Journal of Electrical Engineering & Technology*, 14(2), 993–1000.
41. Attia, S. Y., Mohamed, S. G., Barakat, Y. F., Hassan, H. H., & Al Zoubi, W. (2021). Supercapacitor electrode materials: addressing challenges in mechanism and charge storage. *Reviews in Inorganic Chemistry*, 42(1), 53–88.

42. Liao, C., Zuo, Y., Zhang, W., Zhao, J., Tang, B., Tang, A., & Xu, J. (2013). Electrochemical performance of metal-organic framework synthesized by a solvothermal method for supercapacitors. *Russian Journal of Electrochemistry*, 49(10), 983–986.
43. Borysiewicz, M. A., Dou, J. H., Stassen, I., & Dincă, M. (2021). Why conductivity is not always king–physical properties governing the capacitance of 2D metal–organic framework-based EDLC supercapacitor electrodes: A Ni₃(HITP)₂ case study.
44. Ajdari, F. B., Kowsari, E., Shahrak, M. N., Ehsani, A., Kiaei, Z., Torkzaban, & Ramakrishna, S. (2020). A review on the field patents and recent developments over the application of metal organic frameworks (MOFs) in supercapacitors. *Coordination Chemistry Reviews*, 422, 213441.
45. Poizot, P., Gaubicher, J., Renault, S., Dubois, L., Liang, Y., & Yao, Y. (2020). Opportunities and challenges for organic electrodes in electrochemical energy storage. *Chemical Reviews*, 120(14), 6490–6557.
46. Kandambeth, S., Kale, V. S., Shekhah, O., Alshareef, H. N., & Eddaoudi, M. (2021). 2D Covalent-organic framework electrodes for supercapacitors and rechargeable metal-ion batteries. *Advanced Energy Materials*, 2100177.
47. Yuan, S., Pang, S. Y., & Hao, J. (2020). 2D transition metal dichalcogenides, carbides, nitrides, and their applications in supercapacitors and electrocatalytic hydrogen evolution reaction. *Applied Physics Reviews*, 7(2), 021304.
48. Cheng, F., He, C., Shu, D., Chen, H., Zhang, J., Tang, S., & Finlow, D. E. (2011). Preparation of nanocrystalline VN by the melamine reduction of V₂O₅ xerogel and its supercapacitive behaviour. *Materials Chemistry and Physics*, 131(1–2), 268–273.
49. Shaikh, J. S., Shaikh, N. S., Sabale, S., Parveen, N., Patil, S. P., Mishra, Y. K., & Lokhande, C. D. (2021). A phosphorus integrated strategy for supercapacitor: 2D black phosphorus–doped and phosphorus-doped materials. *Materials Today Chemistry*, 21, 100480.
50. Hao, C., Yang, B., Wen, F., Xiang, J., Li, L., Wang, W., & Tian, Y. (2016). Flexible all-solid-state supercapacitors based on liquid-exfoliated black-phosphorus nanoflakes. *Advanced Materials*, 28(16), 3194–3201.
51. Zhou, D., Wang, F., Zhao, X., Yang, J., Lu, H., Lin, L. Y., & Fan, L. Z. (2020). Self-chargeable flexible solid-state supercapacitors for wearable electronics. *ACS Applied Materials &Interfaces*, 12(40), 44883–44891.
52. Li, L., Lou, Z., Chen, D., Jiang, K., Han, W., & Shen, G. (2018). Recent advances in flexible/stretchable supercapacitors for wearable electronics. *Small*, 14(43), 1702829.
53. Heo, J. S., Eom, J., Kim, Y. H., & Park, S. K. (2018). Recent progress of textile-based wearable electronics: A comprehensive review of materials, devices, and applications. *Small*, 14(3), 1703034.
54. Mishra, A., Shetti, N., Basu, S., Reddy, K., & Aminabhavi, T. (2019). Carbon cloth-based hybrid materials as flexible electrochemical supercapacitors. *ChemElectroChem*, 6(23), 5771–5786.
55. Jia, R., Shen, G., Qu, F., & Chen, D. (2020). Flexible on-chip micro-supercapacitors: Efficient power units for wearable electronics. *Energy Storage Materials*, 27, 169–186.
56. Zhang, H., Cao, Y., Chee, M. O. L., Dong, P., Ye, M., & Shen, J. (2019). Recent advances in micro-supercapacitors. *Nanoscale*, 11(13), 5807–5821.
57. Liu, Y., Zhou, H., Zhou, W., Meng, S., Qi, C., Liu, Z., & Kong, T. (2021). Biocompatible, high-performance, wet-adhesive, stretchable all-hydrogel supercapacitor implant based on PANI@ rGO/Mxenes electrode and hydrogel electrolyte. *Advanced Energy Materials*, 11(30), 2101329.
58. Park, S. H., Goodall, G., & Kim, W. S. (2020). Perspective on 3D-designed micro-supercapacitors. *Materials & Design*, 193, 108797.

59. Li, Y., Xie, H., Li, J., Bando, Y., Yamauchi, Y., & Henzie, J. (2019). High performance nanoporous carbon microsupercapacitors generated by a solvent-free MOF-CVD method. *Carbon*, 152, 688–696.

60. Yang, Y., He, L., Tang, C., Hu, P., Hong, X., Yan, M., & Mai, L. (2016). Improved conductivity and capacitance of interdigital carbon microelectrodes through integration with carbon nanotubes for micro-supercapacitors. *Nano Research*, 9(8), 2510–2519.

61. Xiong, G., Meng, C., Reifenberger, R. G., Irazoqui, P. P., & Fisher, T. S. (2014). A review of graphene-based electrochemical microsupercapacitors. *Electroanalysis*, 26(1), 30–51.

62. Lee, J., Seok, J. Y., Son, S., Yang, M., & Kang, B. (2017). High-energy, flexible microsupercapacitors by one-step laser fabrication of a self-generated nanoporous metal/oxide electrode. *Journal of Materials Chemistry A*, 5(47), 24585–24593.

63. Kurra, N., Hota, M. K., & Alshareef, H. N. (2015). Conducting polymer micro-supercapacitors for flexible energy storage and Ac line-filtering. *Nano Energy*, 13, 500–508.

64. Wang, H., Zhu, C., Chao, D., Yan, Q., & Fan, H. J. (2017). Nonaqueous hybrid lithium-ion and sodium-ion capacitors. *Advanced Materials*, 29(46), 1702093.

65. Yin, J., Zhang, W., Alhebshi, N. A., Salah, N., & Alshareef, H. N. (2021). Electrochemical zinc ion capacitors: Fundamentals, materials, and systems. *Advanced Energy Materials*, 11(21), 2100201.

66. Mohammed, M. M., Abd-Elrahim, A. G., & Chun, D. M. (2020). One-step deposition of a $Ni(OH)_2$-graphene hybrid prepared by vacuum kinetic spray for high energy density hybrid supercapacitor. *Materials Chemistry and Physics*, 244, 122701.

67. Manivel, A., Asiri, A. M., Alamry, K. A., Lana-Villarreal, T., & Anandan, S. (2014). Interfacially synthesized PAni-PMo12 hybrid material for supercapacitor applications. *Bulletin of Materials Science*, 37(4), 861–869.

68. Anandhi, P., Jawahar Senthil Kumar, V., & Harikrishnan, S. (2020). Nanocomposites for Supercapacitor Application. *Handbook of Nanomaterials and Nanocomposites for Energy and Environmental Applications*, 1–24.

69. Chatterjee, D. P., & Nandi, A. K. (2021). A review on the recent advances in hybrid supercapacitors. *Journal of Materials Chemistry A*, 9(29), 15880–15918.

70. Buzzoni, L., & Pede, G. (2012, October). New prospects for public transport electrification. In *2012 Electrical Systems for Aircraft, Railway and Ship Propulsion* (pp. 1–5). IEEE.

71. Song, R., Jin, H., Li, X., Fei, L., Zhao, Y., Huang, H., & Chai, Y. (2015). A rectification-free piezo-supercapacitor with a polyvinylidene fluoride separator and functionalized carbon cloth electrodes. *Journal of Materials Chemistry A*, 3(29), 14963–14970.

72. Ramadoss, A., Saravanakumar, B., Lee, S. W., Kim, Y. S., Kim, S. J., & Wang, Z. L. (2015). Piezoelectric-driven self-charging supercapacitor power cell. *Acs Nano*, 9(4), 4337–4345.

73. Wang, R., Yao, M., & Niu, Z. (2020). Smart supercapacitors from materials to devices. *InfoMat*, 2(1), 113–125.

74. Deng, J., Zhang, Y., Zhao, Y., Chen, P., Cheng, X., & Peng, H. (2015). A shape-memory supercapacitor fiber. *Angewandte Chemie International Edition*, 54(51), 15419–15423.

75. King, P. J., Higgins, T. M., De, S., Nicoloso, N., & Coleman, J. N. (2012). Percolation effects in supercapacitors with thin, transparent carbon nanotube electrodes. *Acs Nano*, 6(2), 1732–1741.

76. Liu, Y. H., Xu, J. L., Gao, X., Sun, Y. L., Lv, J. J., Shen, S., & Wang, S. D. (2017). Freestanding transparent metallic network based ultrathin, foldable and designable supercapacitors. *Energy & Environmental Science*, 10(12), 2534–2543.

77. Yin, L., Chen, Y., Zhao, X., Hou, B., & Cao, B. (2016). 3-Dimensional hierarchical porous activated carbon derived from coconut fibers with high-rate performance for symmetric supercapacitors. *Materials & Design*, 111, 44–50.

78. Wang, X., Xu, W., Chatterjee, P., Lv, C., Popovich, J., Song, Z., & Jiang, H. (2016). Food-Materials-Based Edible Supercapacitors. *Advanced Materials Technologies*, 1(3), 1600059.

79. Yu, J., Zhou, J., Yao, P., Huang, J., Sun, W., Zhu, C., & Xu, J. (2019). A stretchable high performance all-in-one fiber supercapacitor. *Journal of Power Sources*, 440, 227150.

80. Pollet, B. G., Staffell, I., & Shang, J. L. (2012). Current status of hybrid, battery and fuel cell electric vehicles: From electrochemistry to market prospects. *Electrochimica Acta*, 84, 235–249.

81. Buller, S., Karden, E., Kok, D., & De Doncker, R. (2001). IEEE Industry Applications Conference, 36th IAS Annual Meeting (IEEE, 2001), Cat. No. 01CH37248.

82. Ciszewski, M., Koszorek, A., Radko, T., Szatkowski, P., & Janas, D. (2019). Review of the selected carbon-based materials for symmetric supercapacitor application. *Journal of Electronic Materials*, 48(2), 717–744.

83. Cheng, Q., Tang, J., Ma, J., Zhang, H., Shinya, N., & Qin, L. C. (2011). Graphene and carbon nanotube composite electrodes for supercapacitors with ultra-high energy density. *Physical Chemistry Chemical Physics*, 13(39), 17615–17624.

84. Zhang, L. L., Zhou, R., & Zhao, X. S. (2010). Graphene-based materials as supercapacitor electrodes. *Journal of Materials Chemistry*, 20(29), 5983–5992.

85. Dong, Y., Wu, Z. S., Ren, W., Cheng, H. M., & Bao, X. (2017). Graphene: A promising 2D material for electrochemical energy storage. *Science Bulletin*, 62(10), 724–740.

86. Liang, R., Du, Y., Xiao, P., Cheng, J., Yuan, S., Chen, Y., & Chen, J. (2021). Transition metal oxide electrode materials for supercapacitors: A review of recent developments. *Nanomaterials*, 11(5), 1248.

87. Liang, R., Du, Y., Xiao, P., Cheng, J., Yuan, S., Chen, Y., & Chen, J. (2021). Transition metal oxide electrode materials for supercapacitors: A review of recent developments. *Nanomaterials*, 11(5), 1248.

88. Murali, G., Rawal, J., Modigunta, J. K. R., Park, Y. H., Lee, J. H., Lee, S. Y., & In, I. (2021). A review on MXenes: New-generation 2D materials for supercapacitors. *Sustainable Energy & Fuels*, 5, 5672–5693.

89. Hu, M., Zhang, H., Hu, T., Fan, B., Wang, X., & Li, Z. (2020). Emerging 2D MXenes for supercapacitors: Status, challenges and prospects. *Chemical Society Reviews*, 49(18), 6666–6693.

90. Zhang, L., Yang, S., Chang, J., Zhao, D., Wang, J., Yang, C., & Cao, B. (2020). A review of redox electrolytes for supercapacitors. *Frontiers in Chemistry*, 8, 413.

91. Iqbal, M. Z., Zakar, S., & Haider, S. S. (2020). Role of aqueous electrolytes on the performance of electrochemical energy storage device. *Journal of Electroanalytical Chemistry*, 858, 113793.

92. Zhong, C., Deng, Y., Hu, W., Qiao, J., Zhang, L., & Zhang, J. (2015). A review of electrolyte materials and compositions for electrochemical supercapacitors. *Chemical Society Reviews*, 44(21), 7484–7539.

93. Pal, B., Yang, S., Ramesh, S., Thangadurai, V., & Jose, R. (2019). Electrolyte selection for supercapacitive devices: A critical review. *Nanoscale Advances*, 1(10), 3807–3835.

94. Yu, H., Wu, J., Fan, L., Hao, S., Lin, J., & Huang, M. (2014). An efficient redox-mediated organic electrolyte for high-energy supercapacitor. *Journal of Power Sources*, 248, 1123–1126.

95. Galimzyanov, R. R., Stakhanova, S. V., Krechetov, I. S., Kalashnik, A. T., Astakhov, M. V., Lisitsin, A. V., & Tabarov, F. S. (2021). Electrolyte mixture based on acetonitrile and ethyl acetate for a wide temperature range performance of the supercapacitors. *Journal of Power Sources*, 495, 229442.

96. Karuppasamy, K., Theerthagiri, J., Vikraman, D., Yim, C. J., Hussain, S., Sharma, R., & Kim, H. S. (2020). Ionic liquid-based electrolytes for energy storage devices: A brief review on their limits and applications. *Polymers*, 12(4), 918.

97. Yu, L., & Chen, G. Z. (2019). Ionic liquid-based electrolytes for supercapacitor and supercapattery. *Frontiers in Chemistry*, 7, 272.

98. Alipoori, S., Mazinani, S., Aboutalebi, S. H., & Sharif, F. (2020). Review of PVA-based gel polymer electrolytes in flexible solid-state supercapacitors: Opportunities and challenges. *Journal of Energy Storage*, 27, 101072.

99. Kumaravel, V., Bartlett, J., & Pillai, S. C. (2021). Solid Electrolytes for High-Temperature Stable Batteries and Supercapacitors. *Advanced Energy Materials*, 11(3), 2002869.

Index